Video Color Correction for Nonlinear Editors

Video Color Correction for Nonlinear Editors

A Step-by-Step Guide

Stuart Blake Jones

Focal Press

An imprint of Elsevier Science

OXFORD AMSTERDAM BOSTON LONDON NEW YORK PARIS
SAN DIEGO SAN FRANCISCO SINGAPORE SYDNEY TOKYO

Focal Press is an imprint of Elsevier Science.

∞ Recognizing the importance of preserving what has been written, Elsevier Science prints its books on acid-free paper whenever possible.

Library of Congress Cataloging-in-Publication Data
Jones, Stuart Blake, 1961–
Video color correction for nonlinear editors : a step-by-step guide / Stuart Blake Jones.
p. cm.
Includes bibliographical references and index.
ISBN 0-240-80515-1 (pbk. : alk. paper)
1. Motion pictures—Editing—Data processing. 2. Video tapes—Editing—Data processing. 3. Color. I. Title.
TR899 .J667 2002
778.5′35—dc21

2002035226

British Library Cataloguing-in-Publication Data
A catalogue record for this book is available from the British Library.

The publisher offers special discounts on bulk orders of this book.
For information, please contact:
Manager of Special Sales
Elsevier Science
200 Wheeler Road
Burlington, MA 01803
Tel: 781-313-4700
Fax: 781-313-4880

For information on all Focal Press publications available, contact our World Wide Web home page at: http://www.focalpress.com

10 9 8 7 6 5 4 3 2 1

Printed in the United States of America

Contents

Preface

During the time in which I have worked as a colorist, I have been constantly confronted with new challenges. Whether it was to change a gloomy afternoon scene into a sunny day or to simply create a particular color effect, each provided a new and exciting learning experience. Being familiar with the techniques and processes of color correction is vital for not only the operator but also the client. If the client is unfamiliar with the process, he or she is likely to come into the session with expectations that may be difficult to fulfill.

Keeping these ideas in mind, we have compiled this book as a comprehensive how-to guide for taking full advantage of the color-correction tools contained within each of the software applications we mention. We also give the reader a basic knowledge of the techniques of color correction. This book is quite unique in that, along with the sample scenarios and tutorials, there are also QuickTime™ movies and Web page examples contained on the accompanying CD-ROM. These examples visually show the reader exactly how the effect or color correction was created.

The book is divided into seven chapters. In Chapter One we cover the basics of color correction along with a small history of the process. Following this, we cover the role of the colorist along with how to create the ideal color-correcting environment. Chapter Two provides an understanding of the video signal and covers everything from what a waveform monitor or vectorscope is to descriptions of the individual video systems and standards. Chapter Three then describes the various recording formats that the operator may run across. Chapter Four covers the various color-correction tools contained in the software applications. Chapter Five takes us through the various forms of color correction. Chapter Six describes seven sample color-correction scenarios, along with hints and tips for color correction, followed by suggestions for the appropriate applications for various types of color correction. Chapter Seven is a series of tutorials that take us through the processes in a step-by-step fashion, with illustrations. Additionally, many of the tutorials contained in this chapter are accompanied by either Web page examples or QuickTime movies on the companion CD-ROM. Finally, we end with appendices that contain additional useful information, along with a comprehensive glossary.

Acknowledgments

We would like to take this opportunity to extend our thanks to those in the industry who extended their cooperation and support during the writing of this book. Although there is a tremendous amount of competition in the post-production business, the general impression is that it's a close-knit family that is always willing to help and share experiences. There is a tremendous amount of knowledge regarding this industry that is passed on from person to person during the course of time. We would like to thank the following people for their help during the production of this book:

Alicia Awbrey
Daniel A. Brown
Andre Brunger
Kristan Chan
Bob Currier
Alice Eubank
Rob Gonsalves
Jerry Hergenreder
Manny Martinez
Marco Paolini
Mike Rockwell
Chas Smith
Romino Tafaro

I would also like to extend a special thanks to Elinor Actipis of Focal Press for her support and assistance throughout the writing of this publication.

I always look forward to receiving your input. You may contact me via the publisher, or at blake@jcon.demon.co.uk, to let me know of any suggestions you may have for future editions or simply to relay to me how helpful this book was to you. This information would be extremely valuable to me during the production of future editions of the publication.

CHAPTER 1

Getting Started

The Basics of Color Correction

The process of color correction can be best defined as the method of selecting the correct balance of color and image density in order to achieve the proper color rendition or to simply create a specific look. In the video world, the person who carries out this process is known as the colorist, and in the film world, as a color timer.

The process of color correction started many years ago within film laboratories, where the process was applied to the color balancing of negative film material for positive print film reproduction, generally intended for projection in movie theaters. Using this method, the color timer, would balance the negative's color by changing the red, green, and blue values in order to ensure a consistent color balance throughout the material on a scene-by-scene basis. Of course, during this time there were also color-correction applications, such as those used within the realm of still photography, where they would be applied during color print reproduction.

In later years, color-correction capabilities would be extended following the introduction of the telecine machine. The telecine machine is a device found in motion picture film laboratories and video post-production facilities. Its function is to transfer motion picture film material into the video domain. There is generally a device connected to the telecine that is known as a color-correction computer. The color-correction computer allows the operator or colorist to perform color correction and record the changes made on a scene-by-scene basis. During this process, the color-correction computer controls the telecine's internal color corrector.

In the telecine process, two forms of color correction are available: primary and secondary color correction. The method of primary color correction is similar to the capabilities available to the film laboratory color timer except that the colorist has the added ability to adjust the overall black, gamma, and gain levels of the image. These parameters may be accessed from the color-correction computer's remote-control panel. Located on this panel is a set of trackballs or control sticks, which allow the colorist to adjust the color balance of the lift, gamma, and gain within the picture. In addition to this, secondary color correction gives the colorist the ability to adjust the hue, saturation, and luminance of certain colors within the scene. For example, assume that the overall color of the scene is correct but the color of the flowers within the picture is wrong. In this situation, secondary color correction could be used to change the color of the flowers without altering the rest of the picture. The color-correction computer would compile a list of the corrections on a scene-by-scene basis. When the color corrected material was played back, the correction information stored in the computer would be viewed in real time and recorded to tape.

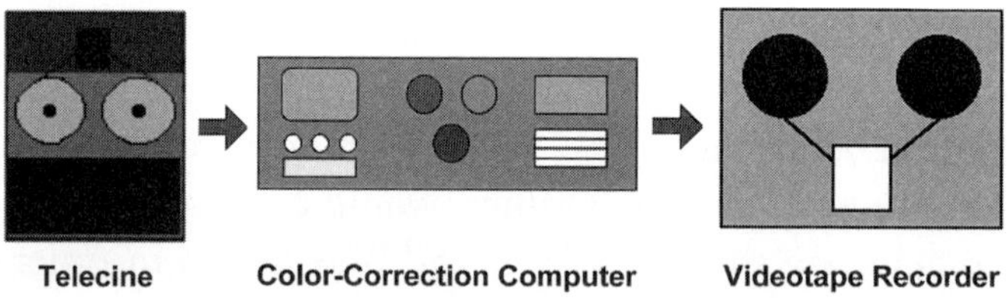

Image 1-1 *Film-to-tape color correction.*

The color-correction computer's capabilities would later be expanded with the introduction of the ability to perform color-correction processing within the computer, as opposed to merely controlling the color-correction electronics within the telecine machine as was done in the past. This addition allowed the colorist the added ability to perform video color correction, or "tape-to-tape" color correction as it's known in the industry. As a result of this change, many postproduction facilities now have dedicated color-correction suites intended solely for performing tape-to-tape color correction. This expansion of the color-correction computer's capabilities caused many postproduction facilities to change their workflow in order to take full advantage of the additional correction power.

Although tape-to-tape color correction does provide added capabilities, it has some definite drawbacks as compared to performing a color correction directly during the telecine transfer. For instance, if the whites are clipped or the black information is crushed during the initial recording, this information cannot be recovered during the tape-to-tape color-correction process. The terms clipped and crushed actually refer to the point when the luminance exceeds 100 IRE (a unit of signal measurement, standing for Institute of Radio Engineers). At this level, the resulting information becomes clipped and therefore lost. On the other hand, when the black information is set too low, the crushed effect results in a loss of detail in the black areas. Additionally, if the material were originally recorded without a sufficient amount of color or chroma information, it would make secondary color correction extremely difficult during a tape-to-tape color-correction session.

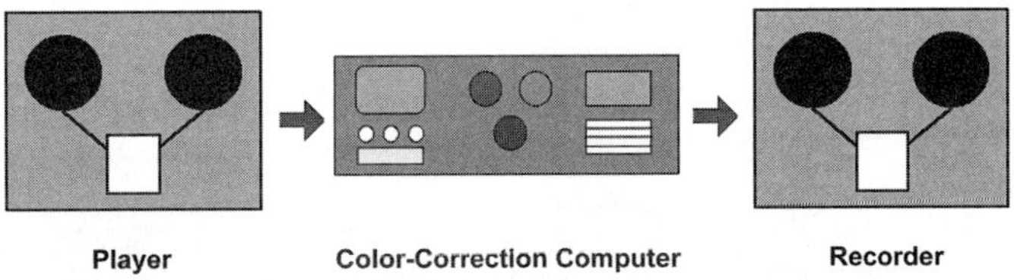

Image 1-2 *Tape-to-tape color correction.*

Color correction during an online editing session was not widely used until recently because most producers preferred to leave the color-correction stage until the project's assembly was completed. Because until recently this was the accepted workflow, any color-correction capabilities contained in an edit suite usually consisted of only very basic color-correction devices, which generally allowed for only primary color correction. Many of these devices found in edit suites were analog color correctors and were used only when absolutely necessary.

The introduction of nonlinear editing systems also brought about the introduction of color-correction tools within their packages. The correction power of these applications could also be expanded with the addition of plug-in software applications. The color-correction capabilities contained in many of the nonlinear editing packages provide the user with the same, and in some cases even more, correction power than would be available to the colorist in a dedicated color-correction suite. These tools have the additional advantage that they can be easily updated in order to accommodate any changes within the industry.

Of course many people ask the same question, "Why do you need to perform color correction?" The answer is quite simple, it can be done to correct for poor weather conditions, to correct for mixed lighting situations, or to simply create a specific effect. Another reason, which can be seen most notably when performing color correction involving a telecine, is to adjust for the difference in contrast ratios. This is essentially the ratio between the light and dark areas of the image. These differences in contrast ratios can be best shown by Image 1-3.

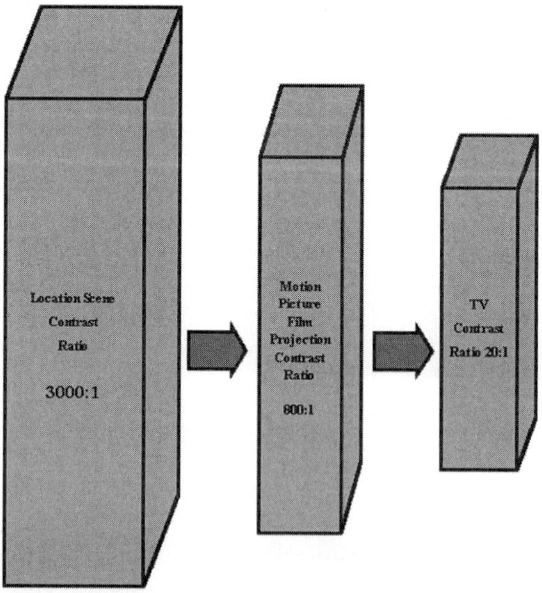

Image 1-3 *A comparison of contrast ratios.*

The Colorist

The colorist's role is indeed a complex one because, in addition to simply performing color correction to balance the color of an image or achieve a look, he or she also has to play the role of advisor in order to give the client and his or her associates information regarding the extent of color correction available. The colorist has the added responsibility to ensure that all parameters of the final product are correct and, in the case of material being prepared for broadcast, that the levels are within "legal" or transmittable limits.

Setting Up Your Color-Correction Environment

The correct working environment for the colorist is extremely important in order to ensure predictable results and a comfortable atmosphere for everyone. It's advisable to eliminate any elements within the room that might influence the outcome of the color-correction process. For instance, a properly calibrated video monitor is essential in order to ensure that the color correction carried out is not compensating for any monitor miscalibration. The monitor used in the color-correction suite should be a broadcast-quality monitor. The use of a consumer-grade monitor is not recommended because the filtering within the monitor may cause inaccurate color rendition in some situations. On the other hand, performing color correction using your computer display is also not ideal because the colors reproduced on the computer monitor will not match the results as seen on a broadcast-quality monitor.

In addition to having a well-calibrated video monitor, it's important to be able to view the video signal using a waveform monitor and vectorscope, whether they are actual scopes or software-generated versions. The use of these scopes allows the operator to observe fine details that may not be visible on the video monitor. The waveform monitor displays a two-dimensional version of the video signal, while the vectorscope shows the hue and saturation of the color.

The utilization of a reference still store is recommended in order to ensure a consistent color balance throughout the correction process. A reference still store contains a library of stills stored during the session, and can be either a device originally intended to be a still store or a digital disk recorder taking only single frames. The library of stills can be referred to at any time to ensure a proper color match and also helps to ensure that the operator's color perception has not drifted during the course of time. The use of digital disk recorders in color-correction suites has become extremely popular in recent times due to their flexibility, since they can be used to store individual frames or full-motion video. Utilizing a digital disk recorder to act as your source machine can be extremely useful because there is little or no waiting time for tape shuttling or cuing.

The workspace in the suite should be large enough to allow freedom of movement for the operator, while allocating a dedicated space in which the client can work and observe the color corrections that are being carried out. The room's color is very important because bright colors can unknowingly affect the colorist's eyes, resulting in a color correction that may subtract from the surrounding wall's color. The use of lighting that is balanced for daylight color temperature is advisable in the proximity of the color-correction area. This form of lighting should also be used whenever reference material is being viewed during a color-correction session. Facilities generally install a daylight-color-balanced tube light into a fixture that allows the light to be directed into a precise area.

Table 1-1. Your Color-Correction Suite Checklist.

Monitor	Ensure that you have well-calibrated broadcast-quality monitor.
Waveform and Vectorscope	In order to monitor the signal properties during the session. They can be in the form of actual scopes or computer applications.
Reference Still Store	Essential to ensure color matching throughout the program material. This can be in the form of either a digital frame store or a digital disk recorder.
Room Lighting	Room lighting balanced for daylight color temperature or "luminant D." This generally takes the form of fluorescent tube lighting.

Summary

The process of color correction has progressed in a relatively short period of time from the basic corrections performed by the film laboratory color timer to the more complex corrections of today's colorist. The introduction of e-cinema, digital film applications, and DTV means that the colorist's role in the industry is constantly growing. In later chapters, we'll see the software packages' various color-correction tools and examine how their efficient use can result in color corrections equal to those achieved in dedicated color-correction suites.

During a color-correction session, a useful operational tip is to expand the magnification of the vectorscope in order to check the color balance of the black information in the picture. Because the center target of the display is black, any deviation from the center will show that there is some color contamination in the darker areas of the picture. However, it should be noted that if the blacks are not balanced but crushed, then a vectorscope could give a false indication. It would give the impression of being properly balanced because the black information would be directly at the center of the display; however, the image displayed on the picture monitor would clearly show that this is not the case.

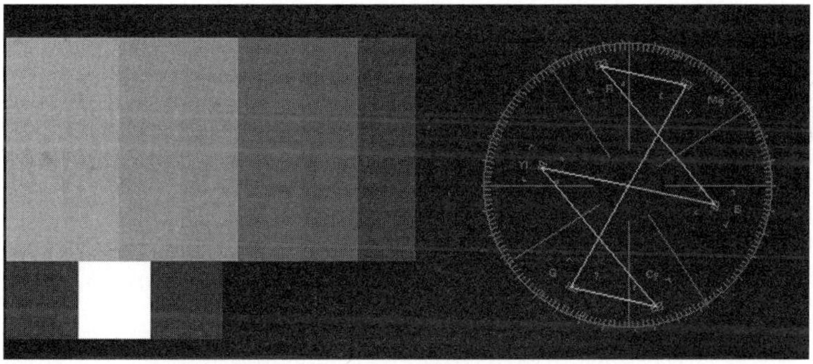

Image 2-2 *The vectorscope display.*

The Parade Display

An alternate display that can be shown on a waveform monitor is known as an RGB or parade display. The parade display shows the red, green, and blue channels individually displayed in sequence. The use of this form of display is the quickest way of viewing the status of each channel's level. For instance, if the colorist notices that the first column, or red channel, is lifted slightly at the baseline, this would indicate a slight amount of red in the dark areas of the picture. Alternately, if the highlights on the third column, or blue channel, were too high, this would indicate a slight amount of blue in the bright areas of the image. Fine adjustments such as these can be carried out with far greater ease using a parade display than with a composite waveform monitor display. Using this type of display is a very rapid way to ensure that the blacks and whites are properly balanced.

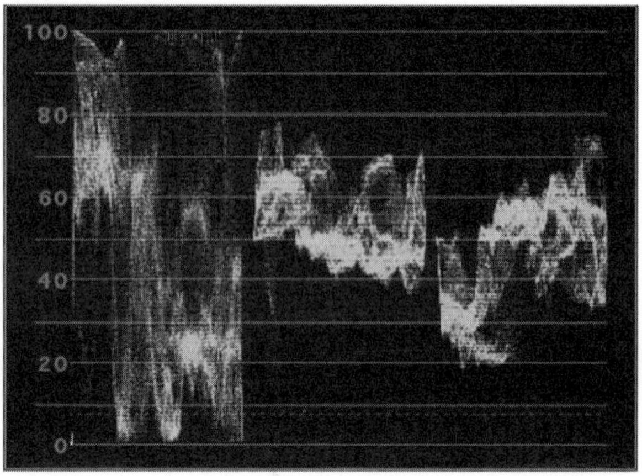

Image 2-3 *The parade display. (Courtesy of Synthetic Aperture)*

The Histogram

The waveform monitor, vectorscope, and parade displays were all originally used in either a conventional editing, color-correction, or telecine suite. There is one display that is quite different and that's the histogram. The histogram display gives the operator the ability to view the image's dynamic range and contrast properties. The display is broken down into an X- and Y-axis in which the X-axis defines the pixel's luminance and the Y-axis shows the quantity of pixels at a level of luminance. The histogram has a measurement scale from 0 to 255, with 0 being black and 255 being white. It's important to note that the histogram does not identify colors on the display but rather shows the amount of light they reflect, and adjusting the histogram can vary the degree of reflection.

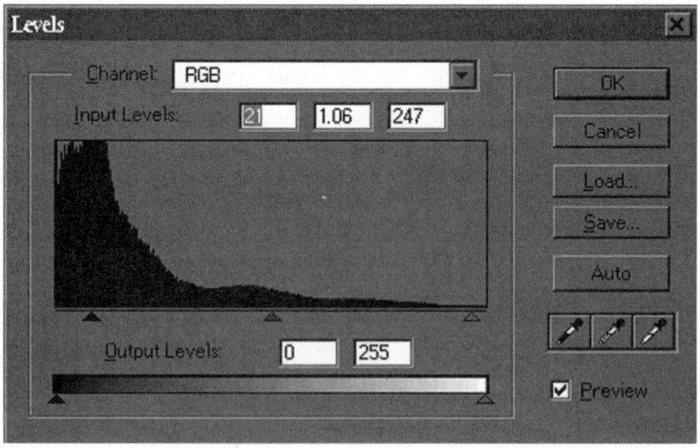

Image 2-4 *Viewing the histogram display.*

Video Standards

NTSC

The NTSC standard, which stands for National Television Standards Committee, is used in the United States, Canada, Japan, and some parts of South America. It's a 525-scan-line interlace system that runs at 60 fields per second. The NTSC standard was initially released in 1941 and first seen only in black and white. The color version was eventually introduced around 1955. It's important to note that the "set-up," which is the reference point where black sits on the waveform monitor, is set at 7.5 IRE in the United States. However, in Japan, the black reference point is set at 0 IRE when viewed on the waveform monitor.

Although NTSC has the advantage of reduced flicker because of the 60 fields or 30 frames per second standard, it also has numerous disadvantages. First of all, the picture's clarity is diminished because of the reduced number of scanning lines. The lower luminance bandwidth results in numerous artifacts within the picture such as a moiré pattern. Additionally, the gamma ratio in NTSC is only 2.2, compared to PAL's 2.8 gamma ratio. Of course, there's also the fact that NTSC is susceptible to slight variations in hue changes.

PAL

The PAL standard is used through most of Europe and Australia. It is a 625-scan-line interlace system. The term PAL stands for Phase Alternate Line because it utilizes a phase alternate line technique during the encoding process to achieve superior color rendition during transmission. The standard is a dual-field interlace format that runs at 50 fields per second. Although the additional scanning lines mean that there is significantly better resolution, having 50 fields per second as opposed to NTSC's 60 fields per second tends to result in a slight image flicker. The PAL standard definitely has some distinct advantages over NTSC. The primary advantage is the additional resolution attained from the additional scanning lines. PAL also has a larger luminance signal bandwidth than NTSC and a higher gamma ratio of 2.8 compared to NTSC's 2.2. It should be noted that the PAL standard also has some disadvantages, the first one being image flicker. Another disadvantage is a diminished signal-to-noise ratio because of the higher bandwidth.

HDTV

Prior to 1995, HDTV (high-definition television) was a maze of different standards developed in various countries throughout the globe. It was soon realized that something had to be done in order to set the standard for HDTV. Because of this, the ATSC, or Advanced Television Systems Committee, was established.

The HDTV standards, which were set up in 1995 by the Advanced Television Systems Committee, basically ended up with four different variations. Some of these variations are in progressive and some in interlace scanning formats. The term "progressive scanning" means that each video frame is actually one field of full resolution, whereas "interlace scanning" means that one video frame is made up of two fields.

Table 2-1. The Various Forms of Standard and High-Definition Television as Laid Out by the ATSC.

Resolution	*Aspect Ratio*	*Frame Rate—Progressive or Interlace*
1920 × 1080	16 × 9	Interlace: 30 Progressive: 24, 30
1280 × 720	16 × 9	Progressive: 60, 30, 24
720 × 483 (SDTV)	16 × 9	Progressive: 60, 30, 24
640 × 480 (SDTV)	4 × 3	Interlace: 30

Video Systems

Component Video

The component video system is composed of three separate signals of red, green, and blue, or RGB, components. A variation is the Y, R-Y, B-Y method used by some manufacturers. In this variation, "R" stands for red and "B" for blue. The "Y" pertains to the luminance information, and the green information is generated from the other components. Both of these methods for maintaining isolated signals yield extremely pure colors and excellent sharpness. Color correction of component video information gives excellent results because of the exceptional purity of the color information.

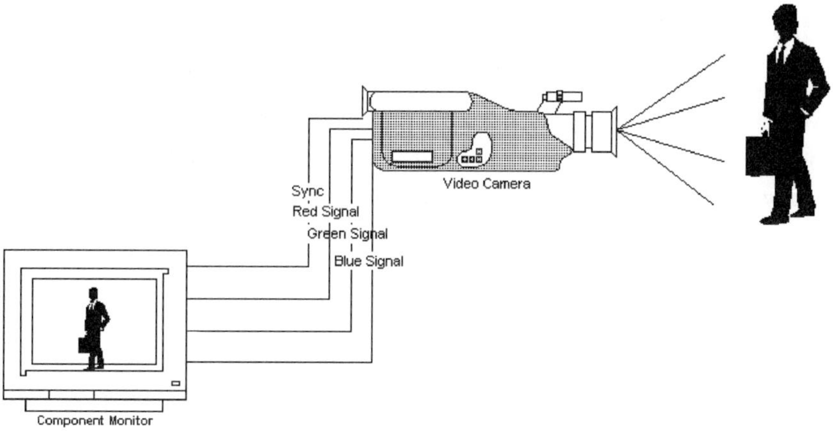

Image 2-5 *A diagram of a component video system.*

Composite Video

In the composite video system, the red, green, and blue signals, along with the luminance and sync information, are combined by means of a composite video encoder that generates a signal in NTSC, PAL, or SECAM. Although a certain amount of quality is lost, this method allows for greater ease of signal handling during transmission. Artifacts that are generated during the encoding process can range from bleeding of color information to moiré to loss of resolution. Unfortunately, any artifacts generated during the encoding process can create complications during the color-correction process.

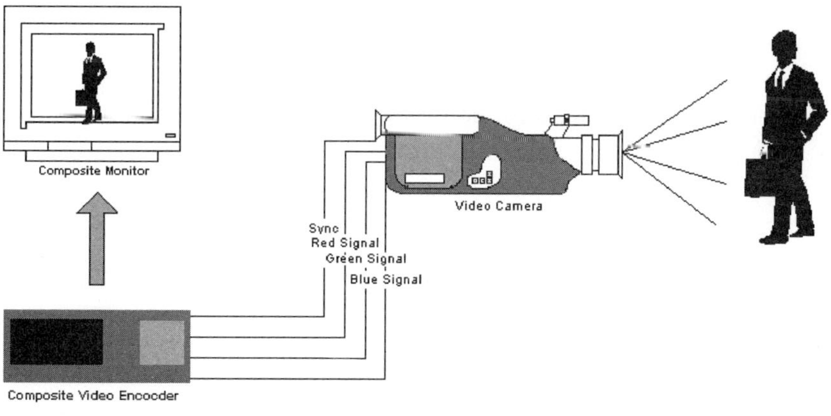

Image 2-6 *A diagram of a composite video system.*

Analog Video

Analog video is essentially the representation of information by continuous voltage variations. An analog signal can be generated in a component or composite format. There is also an alternate format used in analog video that is somewhat in between analog component and composite video. This format is known as S-Video. The format consists of two individual signals, one used for composite chroma and the other for luminance. When capturing video using a PC capture card, S-Video can supply extremely good results.

Digital Video

In digital video, the analog values are represented in the form of binary numbers. When the analog values are measured, this is known as sampling. A higher frequency of sampling results in superior image detail within the signal. Analog signals can be converted to a digital format with an analog-to-digital converter. Alternately, the reverse process can be carried out by using a digital-to-analog converter, or D/A. There are several forms of digital video, as shown in Table 2-2.

Table 2-2. The Various Forms of Digital Video.

4:4:4	This form is seen generally in high-end postproduction systems and contains full and equal amounts of chrominance and luminance information. The first number represents the luminance channel, the second the R-Y chrominance channel, and the last the B-Y chrominance channel.
4:2:2	This form is widely used in television postproduction and is recorded on machines such as D-1 videotape recorders and digital Betacam® videorecorders. The terminology stands for full-resolution luminance and only half the amount of chrominance resolution.
4:1:1	4:1:1 is used in NTSC DV, DVCAM, and DVCPRO. Using this method, the luminance information is at full resolution; however, the chrominance resolution is half of what it is in 4:2:2.
4:2:0	This form is used for PAL DV and DVD. As in 4:2:2, there is full-resolution luminance; however, the chrominance only has half resolution on every other scanning line.

CHAPTER 3

Videotape Recording Formats

During the course of work, the operator generally has little way of knowing what type of source material the client will appear with. It could be anything from a DV videotape right up to a D-6 videotape. Of course, the source tape will greatly affect the quality of color correction available. The remainder of this chapter has a brief explanation of the various formats available.

Analog Recording Formats

In many cases, analog recording formats can be used for color correction and achieve acceptable results. There are two types of analog recording formats: composite and component. For color-correction purposes, the best type to use is a component format. Attempting to color correct from a composite format can result in noise magnification generated during the encoding process. Additionally, color bleeding, which occurs within composite recordings, can make isolating colors for secondary color correction extremely difficult.

1-Inch Type C

The 1-Inch Type C format is an open-reel format that records a composite video signal onto a 1-inch-wide videotape. There are three audio tracks, two for the program audio and one for longitudinal time code. This format was introduced in 1978 and has been widely used throughout the production and postproduction industry. The maximum playing time available is 180 minutes, and the tape moves through the machine at a speed of 9.6 inches per second.

8mm

The 8-mm format is a composite analog format. It was originally released by Eastman Kodak. It provides an image quality that is slightly superior to that of VHS. It makes the recording on an 8-mm metal-particle videotape enclosed within a cassette. It supports one stereo AFM and one stereo PCM audio track and has a maximum recording time in NTSC of 240 minutes, running at LP speed, using a P6–120 videocassette. In PAL, the format has a maximum recording time of 180 minutes, using a P5–90 videocassette, running at LP speed.

Betacam

Betacam® is an analog component format developed by Sony Corporation in 1982 and uses a ½-inch (12.7-mm) videotape enclosed within a cassette. The format supports a maximum recording time of 30 minutes. Besides being able to record composite video, the format has the option of recording a Y, R-Y, B-Y component signal and supports up to two channels of audio and a time-code track. This format has become extremely popular for electronic news gathering (ENG) applications.

Betacam SP

Sony Corporation released the Betacam SP® format in 1987. The "SP" stands for superior performance. The format supports videocassettes with a longer running time than Betacam. As with its Betacam counterpart, Betacam SP uses a ½-inch (12.7-mm)-wide videotape enclosed within a cassette. The format supports a maximum recording time of 90 minutes and has four channels of audio, two of which are FM tracks, which provide high-quality audio reproduction. The change from Betacam to Betacam SP saw the format evolve from being used for news-gathering applications and industrial use to being widely used within the postproduction industry.

Betamax

Betamax® is an analog videocassette format intended for consumer use that was originally developed by Sony. It utilizes a half-inch (12.5-mm) ferro-chrome videotape enclosed within a cassette. It supports up to two channels of audio and has a maximum recording time of 300 minutes in NTSC. The maximum recording time in PAL is 215 minutes.

Hi-8

The Hi-8® format is an analog format that can receive a video input in composite or Y/C mode, which separates the chrominance and luminance information. It was introduced by Sony in 1989. It records the information onto an 8-mm-wide metal-evaporated cassette-enclosed videotape. It is basically a consumer format; however, it is sometimes used for professional applications. The format has a maximum recording time of 240 minutes in NTSC, running at LP speed, using a P6–120ME videocassette. In PAL, the maximum recording time is 180 minutes in LP mode, using a P5–90ME videocassette.

S-VHS

The S-VHS format utilizes a special high-energy videotape and provides superior resolution over its VHS counterpart by receiving the input video in Y/C mode, which separates the chrominance from the luminance information. The use of the Y/C input signal significantly improves the color reproduction. The format has a maximum recording time of 480 minutes, using an ST-160 videocassette running at EP speed in NTSC. The maximum recording time in PAL is 480 minutes at LP speed, using an SE-240 videocassette.

U-Matic

The U-Matic®, or ¾-inch format as it's sometimes known, is essentially an industrial format released in 1968 and has been widely used throughout the production and postproduction industry due to its compact size and quality. It was widely used within the news business for field ENG. It records using a ¾-inch-wide videotape enclosed within a cassette and has a maximum recording time of 75 minutes. The U-Matic format is a composite analog video format that has two audio tracks, one of which is frequently used for the time code. In order to maintain superior color rendition, the chrominance and luminance information is separated during recording and restored during playback. There are also certain professional models that have their own dedicated time-code channel, which is known as "address track time code."

VHS

This is a consumer format that records an analog composite signal along with two channels of audio. The format records using a ½-inch-wide videotape enclosed within a specially designed cassette. There are three speeds available. These are: 1.94 inches per second (known as SP or standard play), which yields the best results; LP (long play) mode, which records at 0.66 inches per second; and finally, SLP (super long play), which records at 0.44 inches per second. The best quality is of course achieved when using the SP mode. The format has a maximum recording time of 320 minutes in LP mode, using a T-160 videocassette in NTSC, and 600 minutes using an E-300 in PAL at LP mode.

Digital Recording Formats

Performing color correction from digital recording formats is by far the best route to take if possible. There are two types of digital formats: composite digital and component digital. For color-correction purposes the ideal format is component digital because it allows for easy isolation of colors during secondary color correction. Care should always be taken when dealing with formats that contain any sort of compression. For example, D-1 is by far the best choice because it involves no compression at all. Even though a format such as Digital Betacam® has a compression ratio of two to one, this level of compression does not create difficulties in most cases. However, performing secondary color correction using formats such as HDCAM, which has a compression ratio of 4.4 to 1, can occasionally cause compression artifacts to become visible. Of course, the best sampling ratio to work with for color-correction purposes is 4:2:2; however, acceptable results can be achieved with 4:1:1 or 4:2:0.

Betacam SX

Sony released the Betacam SX® format in 1996. It's a component digital format that has a sampling ratio of 4:2:2 and a 10:1 MPEG 2 compression ratio. The recording is done on a ½-inch metal-particle videotape enclosed within a cassette. The format supports up to four digital audio tracks plus a time code. The maximum recording time for Betacam SX is 184 minutes, and the tape moves through the transport at a play speed of 59.6 millimeters per second, or 2.3 inches per second.

D-1

The D-1 4:2:2 component digital format was introduced in 1986 and has since become the industry standard for uncompressed digital video recording. The recording is made on a ¾-inch-wide videotape enclosed within a cassette. The maximum recording time for the format is 90 minutes. The format supports up to four digital audio channels plus a cue track. For color correction of standard definition material, the D-1 format is by far the best format to use because it represents the highest quality available and has no video compression.

D-2

The D-2 format is a composite digital format that records information onto a ¾-inch-wide videotape enclosed within a cassette and runs at a speed of 5.185 inches per second, or 131.7 millimeters per second. The D-2 format has four digital audio channels and supports a maximum recording time of 126 minutes.

D-3

The D-3 format is a composite digital format developed by the NHK in 1991. It records the information onto a ½-inch-wide videotape enclosed within a cassette at a speed of 3.28 inches per second, or 83.88 millimeters per second. It has a maximum recording time of 246 minutes. It supports up to four channels of digital audio and has one channel allocated for the time code. It's a format that is widely used throughout the broadcast industry.

D-5

The D-5 4:2:2 component digital format was released in 1994 by the Panasonic corporation. The format uses the same type of cassette as the D-3 format, which uses a ½-inch-wide videotape enclosed within a cassette. The format also has four digital audio channels and a dedicated time-code channel. The maximum playing time for D-5 is 124 minutes. There is a high-definition television (HDTV) adapter, which can be added in order to permit HDTV recordings with a four-to-one compression ratio.

D-6

D-6 is a compression-free, component digital HDTV recording format developed by Philips and Toshiba. The recording is made onto a ¾-inch-wide videotape enclosed within a cassette, providing a maximum recording time of 64 minutes. It also has twelve channels of audio along with a time-code channel.

D-7 (DVCPRO)

The D-7 format is also known as DVCPRO and was developed by the Panasonic Corporation. The format was developed for use within industrial and broadcast applications, generally intended for ENG applications. It records the information using a special MP or metal-particle 6-mm videotape with a maximum recording time of 184 minutes. The format uses a digital sampling ratio of 4:1:1, which means that the luminance information is at full resolution; however, the chrominance resolution is half of what it is in 4:2:2. This is a component digital format that has a compression ratio of five to one and also has two digital audio tracks.

DCT

The DCT® format is a 4:2:2 component digital format. The "DCT" stands for Digital Component Technology® and was developed by the Ampex Corporation. The format records onto a ¾-inch- or 19-mm-wide videotape enclosed within a cassette. The format has a compression ratio of two to one and supports up to four channels of audio and a time-code channel. The DCT format has a maximum recording time of 208 minutes.

Digital Betacam

In 1995, Sony released the Digital Betacam® format. As with Betacam SP, Digital Betacam uses a ½-inch (12.7-mm) videotape enclosed within a cassette that is specially marked for use within a Digital Betacam machine. It is a 4:2:2 component digital format with a two-to-one compression ratio, along with four digital audio tracks and a time-code track. The maximum playing time is 124 minutes. This format is widely used throughout the industry and provides excellent results for color correction.

Digital S (D-9)

The Digital S format, also known as D-9, was developed by the JVC Corporation and is a component digital format. It has a sampling ratio of 4:2:2 and the recording is made on a ½-inch metal-particle videotape enclosed within a cassette. It actually uses the same cassette shell that is used for the VHS format. The format supports up to four digital audio tracks and the tape moves through the transport at a speed of 2.1 inches per second. It has a compression ratio of 3.3:1 and a maximum recording time of 104 minutes.

DV/DVC

The DV format, which was also referred to as DVC, is a component digital recording format that has a five-to-one compression ratio. It also supports up to two channels of digital audio. The information is recorded on a special 6-mm metal-evaporated videotape enclosed within a cassette. The format has a maximum recording time of 6.9 hours, using the long play mode with a DVCAM® 184-minute videocassette. It has a sampling ratio of 4:1:1 in NTSC and 4:2:0 in PAL. It is widely used for consumer and professional applications.

DVCAM

The DVCAM format from Sony is a component digital recording format. It supports up to two channels of digital audio and has a maximum recording time of 184 minutes, recording on a special 6-mm metal-evaporated cassette-enclosed videotape. The format is generally intended for professional or industrial applications. In NTSC, it records with a sampling ratio of 4:1:1, which means that the luminance information is at full resolution; however, the chrominance resolution is half of what it is in 4:2:2. When used in PAL, the sampling ratio is 4:2:0, which means that there is full luminance resolution as in 4:2:2; however, the chrominance only has half resolution on every other scanning line.

DVCPRO 50

The DVCPRO 50 format is a component digital format that uses a sampling ratio of 4:2:2 and has a compression ratio of 3.3:1. It has a maximum recording time of 90 minutes and supports up to four channels of digital audio. The information is recorded on a cassette-enclosed 6.35-mm-wide videotape.

Digital 8

The Digital 8® format from Sony is a component digital format that has a sampling ratio of 4:1:1 in NTSC and 4:2:0 in PAL. It uses a compression ratio of five to one. The format records the information onto an 8-mm-wide videotape enclosed within a cassette running at a speed of 28.666 millimeters per second, or 1.12 inches per second. The maximum recording time is 90 minutes on either a metal-evaporated or metal-particle videotape, and it supports up to two channels of digital audio.

HDCAM

This is an HDTV format that records a digital HDTV signal plus four channels of audio onto a ½-inch-wide videotape enclosed within a cassette, with a compression ratio of 4.4 to 1. The HDCAM format was introduced in 1997 and is basically to HDTV what Digital Betacam is to standard-definition television (SDTV). The maximum recording time for HDCAM is 124 minutes.

CHAPTER 4

The Color-Correction Tools

Adobe Systems, Inc.

Adobe Systems, Inc. produces a wide range of software applications. Both Adobe After Effects® and Adobe Premiere® contain a huge selection of color-correction tools that can easily handle even the most complicated color-correction scenarios.

Adobe After Effects

Adobe After Effects version 5.5 contains a vast array of extremely powerful color-correction tools, combined with Adobe's ability to import and export between applications. The operator can easily create colored vignettes or even apply color-correction effects in specially defined areas of the image by utilizing the mask function. Additionally, Adobe After Effects can also accommodate plug-in applications that further expand the system's color-correction power. Following are descriptions of the various color-correction tools contained within Adobe After Effects.

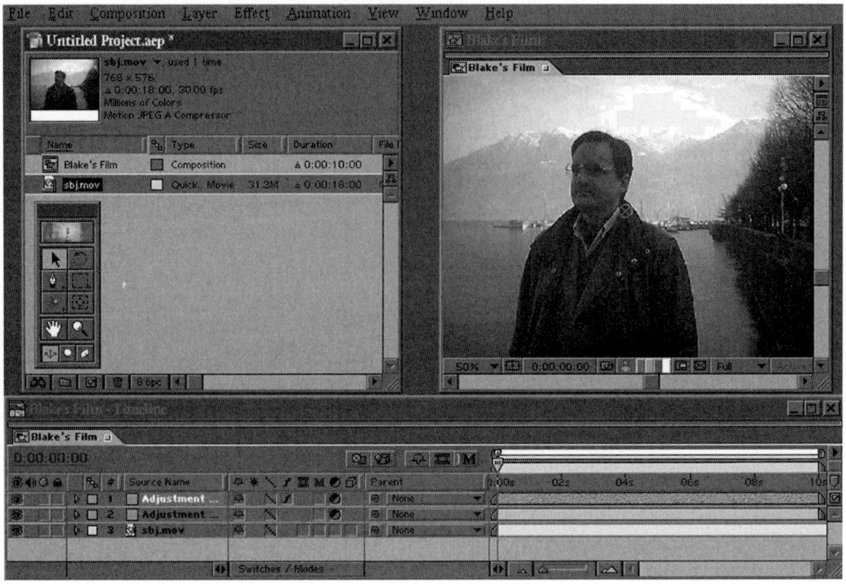

Image 4-1 *The Adobe After Effects desktop.*

Brightness and Contrast

This filter provides individual slider controls for the overall adjustment of the image's brightness and contrast. On the upper portion of the control window is also a reset button, which can reset the controls to their neutral position. The Brightness and Contrast filter is extremely effective as a starting point for your primary color correction.

Color Balance

This effect allows the user to adjust the overall color balance of the image by adjusting the sliders dedicated to the shadows, mid-tones and highlights of either red, green, or blue. These adjustments can provide extremely effective results for primary color-correction applications.

Channel Mixer

The channel mixer provides individual slider controls for adjusting the color by using a mixture of the color channels. This tool can be extremely useful for creating specialized effects and even color washes.

Curves

The Curves effect allows the operator to adjust the image's tonal values by adjusting the Curves graphic display. The display can be changed either by moving the spines of the curve with your mouse or by drawing the curve with the pen function. The upper-right part of the graph represents the highlights, while the middle part shows the mid-tones, and the lower-left portion displays the shadow areas. The display can adjust the red, green, or blue values independently or together. It also provides the ability to adjust only the alpha channel.

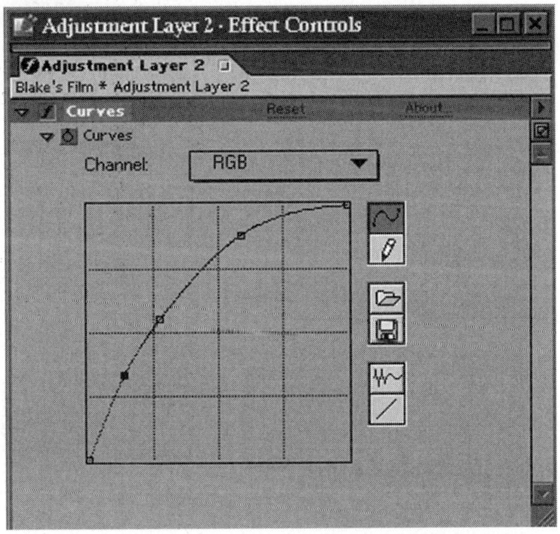

Image 4-2 *The Adobe After Effects Curves effect display.*

Hue/Saturation

This effect can be best described as an extremely powerful secondary color corrector. This tool has a series of fixed vector correction positions for red, green, blue, yellow, cyan, and magenta. There is also a Colorize portion of this effect, which will place a customizable color wash throughout the entire picture. Additionally, within the pull-down menu that selects the preset vectors, there is also the option of selecting Master. When Master is selected, the hue, saturation, and brightness controls will affect the entire image.

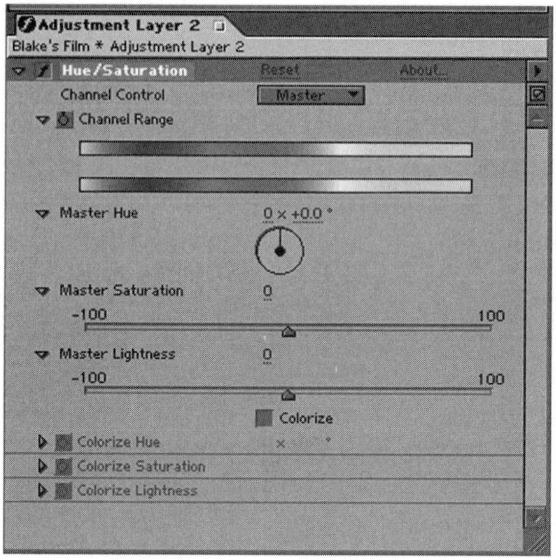

Image 4-3 *The Hue/Saturation effect window in Adobe After Effects.*

Levels

The operator may adjust the contrast and brightness of the image by using the histogram display. In addition to this, the user also has ability to alter the values in the red, green, or blue channels individually. The levels may be adjusted by using the slider controls at the lower portion of the window or by adjusting the histogram display with the mouse.

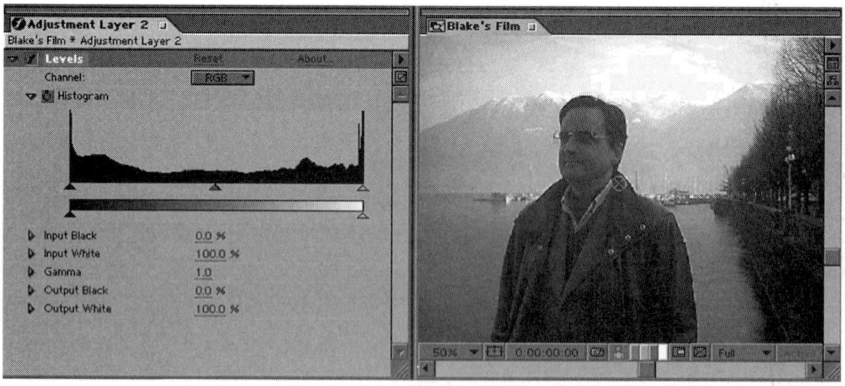

Image 4-4 *The Levels Control window within Adobe After Effects.*

Levels (Individual Controls)

This effect is similar to the Levels effect using the histogram display. The difference here is that there are individual controls for red, green, blue, and alpha channel adjustments. The levels may be adjusted by using the slider controls at the lower portion of the window or by adjusting the histogram display with the mouse.

Change Color

The Change Color effect selects the color to be changed by using the eyedropper, which in turn generates a matte signal. The user may then change the hue, saturation, and luminance of the selected color by adjusting the transform controls. The appropriate controls will vary the information that is defined within the mask. The Match pull-down menu indicates whether the color isolation is being defined by the hue, chroma, or RGB.

Color Balance (HLS)

This effect allows the user to control of the image's hue by using a rotary control along with the dedicated sliders for adjustment of the saturation and luminance, which are located within the control window. These controls can be very useful because their adjustment can provide a tremendous amount of control over the image.

Gamma/Pedestal/Gain

The Gamma/Pedestal/Gain effect provides a control for overall black stretch, along with separate slider controls for adjusting the red, green, and blue values for the pedestal, gain, and gamma. This effect is extremely effective for carrying out primary color correction throughout the whole image.

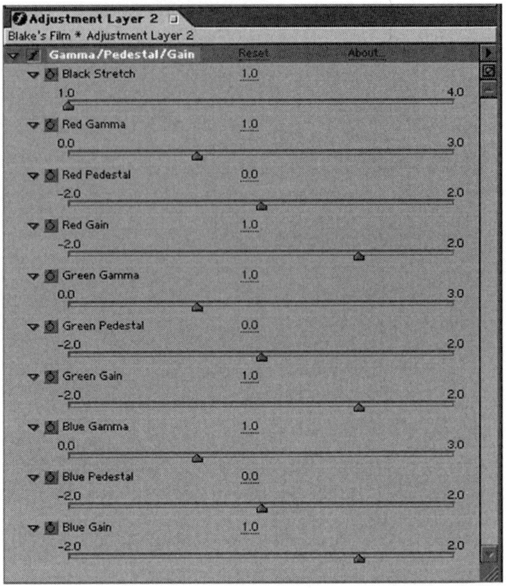

Image 4-5 *Adobe After Effects Gamma/Pedestal/Gain Control window.*

Tint

This tool allows the operator to select a specific color for the blacks and a specific color for the whites. Following this, the operator may then create a blending effect using these values. This is an extremely useful tool for numerous applications, as are used, for example, when creating a dual-tone effect.

Equalize

This effect essentially equalizes the brightness and color information throughout the image. Its operation is very similar to that of its Adobe Photoshop® counterpart.

Leave Color

The Leave Color effect allows the operator to select a color within the scene by using the eyedropper and then to desaturate the rest of the picture, leaving only the selected area in color. This type of effect is extraordinarily useful for special situations such as color correcting commercials and music videos.

Broadcast Colors

The Broadcast Colors effect can be used to legalize colors within an image that may contain illegal levels of color. It can also be used to ensure that the color correction created will be correctly presented on a standard television.

Adobe Premiere

As with Adobe After Effects, Adobe Premiere version 6.0 contains an impressive array of color-correction tools. Actually, many of the tools or effects that are in Adobe Premiere are originally from Adobe After Effects, and they can be easily identified in the effects window because their icon is the cyan-colored plug-in symbol. These color-correction capabilities, combined with Adobe's ability to exchange files between applications, provide great flexibility. For instance a filmstrip from Adobe Premiere can be easily exported to Adobe

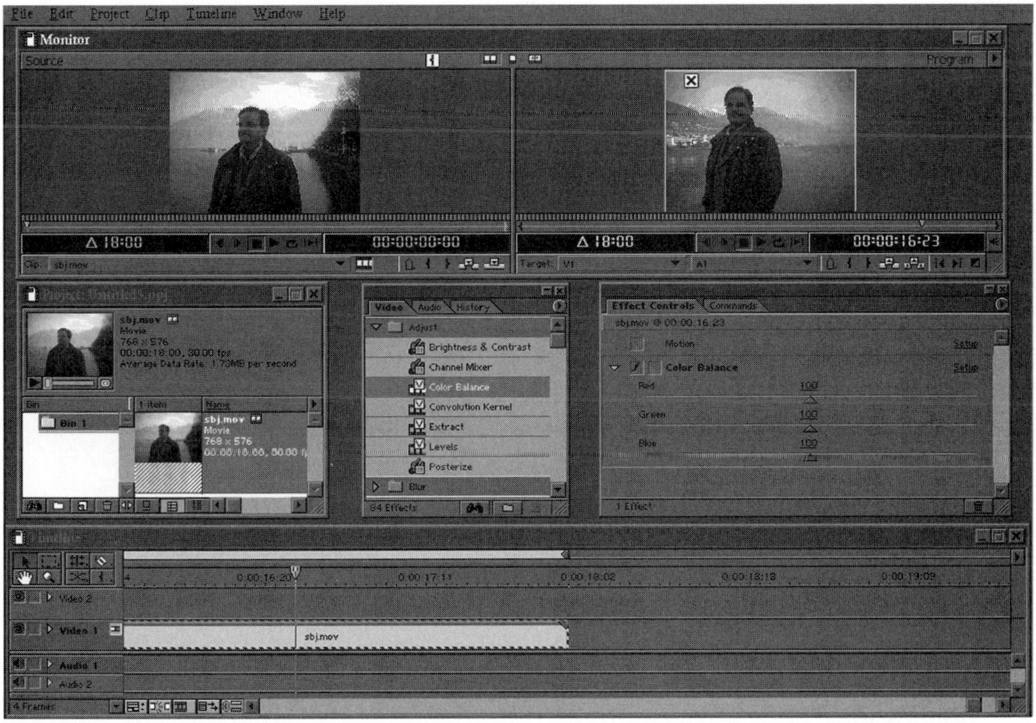

__Image 4-6__ The Adobe Premiere desktop display.

Photoshop, corrected and then reimported into Adobe Premiere. Below is a brief explanation of the tools contained within Adobe Premiere.

Brightness and Contrast

This filter provides individual slider controls for the overall adjustment of the image's brightness and contrast. On the upper portion of the control window is a reset button, which can be used to reset the controls to their neutral position. The Brightness and Contrast filter is very effective as a starting point for your primary color correction.

Channel Mixer

The channel mixer provides individual slider controls for adjusting the color using a mixture of the color channels. This effect can be extremely useful for creating specialized effects and even color washes.

Image 4-7 *The Adobe Premiere Effect Library.*

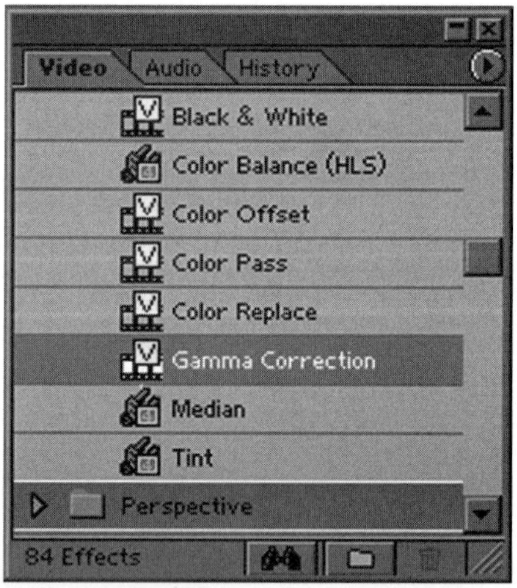

Color Balance

This effect allows the user to adjust the overall color balance of the image by adjusting the three sliders dedicated to red, green, and blue. These adjustments can provide extremely effective results for primary color-correction applications.

Levels

The operator may adjust the contrast and brightness of the image by using the histogram display. The user also has ability to alter the values in the red, green, and blue channels individually using this effect.

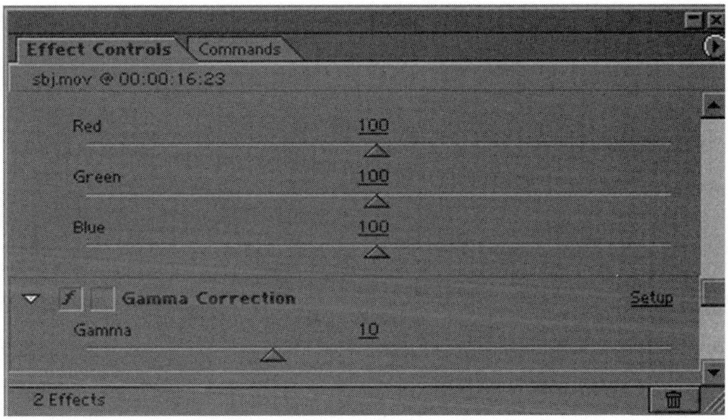

Image 4-8 *The Effect Controls window within Adobe Premiere.*

Black and White

This effect changes the image into black and white or monochrome. This particular effect can be extremely useful when combined with a vignette.

Color Balance (HLS)

The Color Balance (HLS) effect allows the user to alter the hue, saturation, and luminance of the image by using the dedicated slider controls. This can be very effective for carrying out primary color correction throughout the entire image.

Color Pass

Once this effect is selected, the image will be changed into black and white. Using the eyedropper, the operator may select a part of the image that will be in color, while leaving the rest in black and white.

Color Replace

The Color Replace effect allows the user to select a specific color within the image and then select a color to replace it with.

Gamma Correction

This correction tool allows the operator to adjust the image's overall gamma level. If the user clicks with the mouse on the setup label, a preview display opens, along with a slider control for adjusting the image's gamma level.

Tint

This tool allows the operator to select a specific color for the blacks and a specific color for the whites. Following this, the operator may then create a blending effect using these values.

Broadcast Colors

The Broadcast Colors effect can be used to legalize colors within an image that may contain illegal levels of color. It can also be used to ensure that the color correction created will be correctly presented on a standard television.

Apple Computer, Inc.

Final Cut Pro 3

Final Cut Pro 3™ from Apple Computer, Inc. contains an impressive array of color-correction tools. It's important to point out that the color-correction filters within Final Cut Pro™ actually work in a cascade sequence, which is extremely useful for creating a correction with one filter and then applying another correction on top of it to further modify the existing correction. This type of correction power, combined with the other features of the system, allows the operator to easily handle even the most difficult of color-correction applications. Listed below is a brief description of the various color-correction and measuring tools contained within Final Cut Pro.

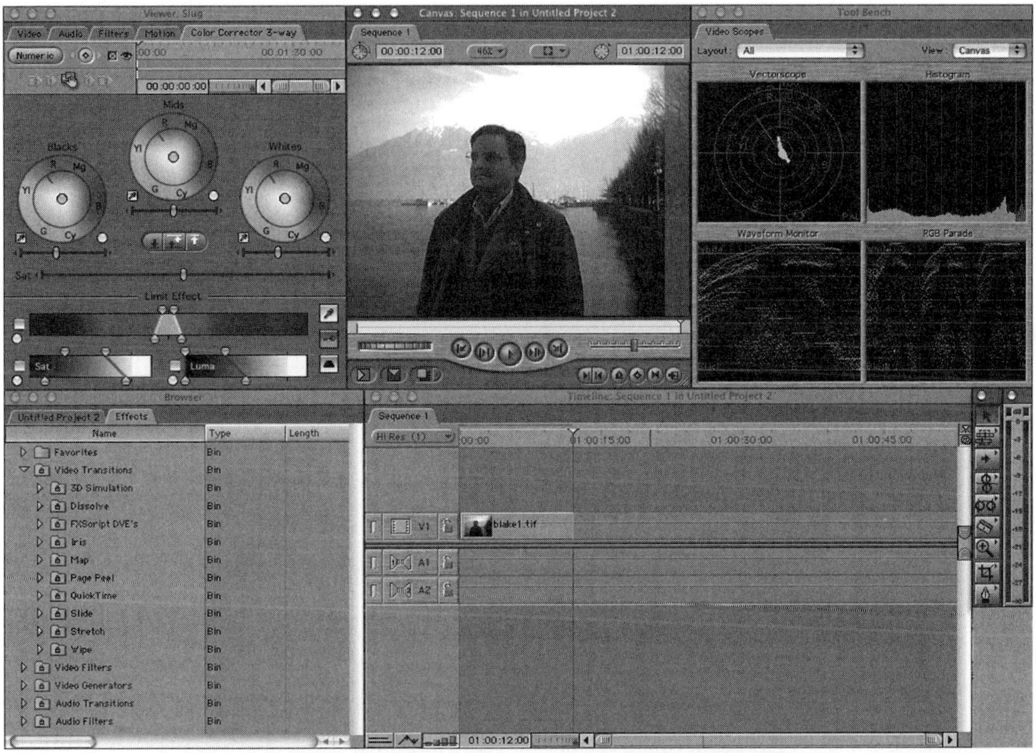

Image 4-9 *The Final Cut Pro 3 Desktop.*

Color- and Brightness-Measuring Tools

Range-Checking Tool

This tool allows the user to view illegal chrominance and luminance information and can be found in Final Cut Pro's View menu. For instance, if the Excess Luma option is enabled, red stripes will appear in the luminance areas that are above 100 percent. The presence of yellow stripes will indicate luminance levels within the 90–100 percent region, and green stripes indicate that the luminance levels are within range. High luminance levels within the image are displayed to the operator by showing an exclamation point. If the Excess Chroma option is enabled, the red stripes will indicate illegal chroma information, and green stripes will, of course, indicate that everything is within range. As with high luminance levels, the presence of an exclamation point in this mode will show that the chroma levels may be a bit too high. If the Both option is selected, the red stripes will indicate luminance and chrominance information that is out of range, and the yellow exclamation mark will indicate levels that are too high.

The Tool Bench Window

The Tool Bench window contains a Video Scope tab, which gives the user instant access to the Waveform Monitor and Vectorscope, along with the Histogram and Parade displays. These displays allow the user to precisely view the signal's properties and view details of the video signal that may not be visible on the viewing monitor.

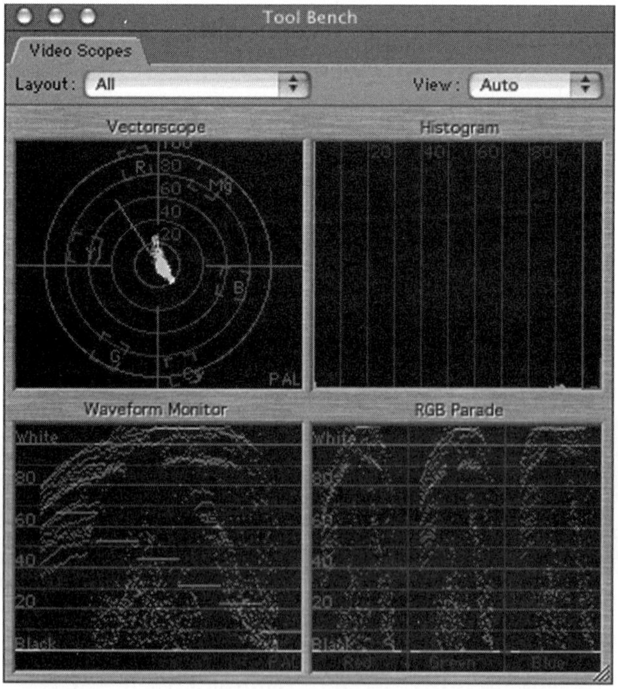

Image 4-10 *The Video Scopes within Final Cut Pro's Tool Bench.*

Color-Correction Filters

Limit Effect

The Limit Effect display, which appears at the lower portion of both the Color Corrector filter window and the Color Corrector 3-Way filter, is essentially an extremely powerful secondary color corrector. To the right of this display are three buttons: the eyedropper, which allows the user to select the color to be modified; the Key button, which permits the viewing of the pre- and postcorrections, along with the matte signal created by the selection; and an Invert control, which allows the inversion of the correction from inside the selected area to outside the selected area. The Color Range control displays the vector that has been selected. The upper portion of the vector, or Top Handle, as it's known, selects the range of colors. The lower portion of the selected vector, or Bottom handle, adjusts the selection tolerance. When the vector is moved from left to right in this display, the hue of the color range selected is changed.

The last two controls are the Saturation and Luminance controls. These two controls allow the user to define the vector's qualification by either the saturation or luminance differences.

Color Corrector

This is a filter that performs basic color correction without the enhanced features of the Color Corrector 3-Way filter. The Color Corrector filter window contains two circular indicators: one for adjusting the color balance of the white information in the image and the other, a circular indicator, for adjusting the overall hue for the image. Underneath these indicators are four slider controls that allow level adjustments of whites, mids, blacks, and saturation. The Color Corrector filter has an extremely handy button called Auto-balance. When this is selected the pointer turns into an eyedropper and the pixel selected will be balanced to create a neutral white balance. This control should not be confused with the Auto White and Black Level buttons, which can also be extremely useful. These controls examine the image information and automatically set the optimum levels of black and white information. There is also an Auto Contrast button, which performs simultaneously the Auto Black and White controls.

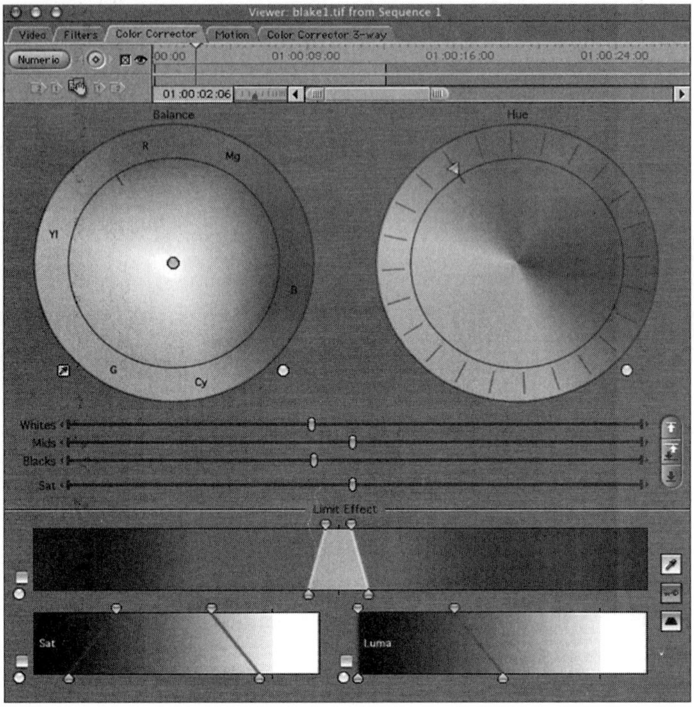

Image 4-11 *The Color Corrector filter display.*

Color Corrector 3-Way Filter

This color corrector provides the user with far more enhanced color-correction capabilities, allowing for individual adjustment for the lift, mid-tone, and luminance information. At first glance, the Color Corrector 3-Way filter display looks quite similar to the Color Corrector filter display. However, the Color Corrector 3-Way color balance controls are laid out in a very similar fashion to that of a standard color-correction computer seen in a color-correction suite. There are separate color balance controls for blacks, mids, and whites, along with slider controls for adjusting the master levels of each. Additionally, there is also one large slider control, which adjusts the overall saturation of the image. Below this is the Limit Effect section, which is virtually identical to that on the Color Corrector filter display. There are also three buttons just above the saturation slider control, which are the Auto Black, Auto White, and Auto Contrast controls. When one of these is selected, it automatically sets the optimum level for either the overall black or white information within the image by analyzing the image properties.

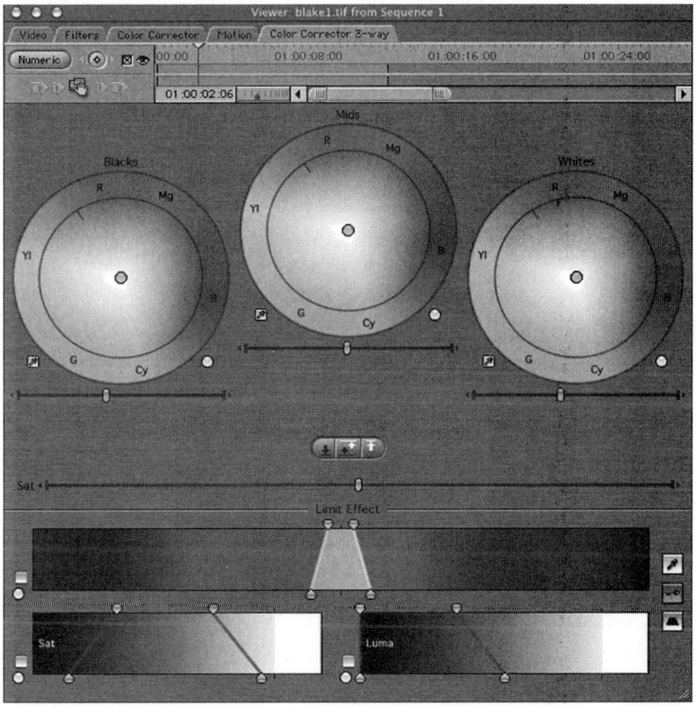

Image 4-12 *The Color Corrector 3-Way filter display.*

Desaturate Highlights or Desaturate Lows

These are two filters that allow the operator to desaturate either highlighted or lowlighted areas in order to eliminate color contamination in these areas. In essence, they are actually one filter, but with different adjustments. Both the Desaturate Highlights and Desaturate Lows controls contain adjustments for the percentage of luminance above or below that of the area to be desaturated (this level is called Begin At). Each filter also contains a Softness control in order to create a smooth transition between the affected and unaffected portion of the image. Finally each filter contains an adjustment called Amount of Desaturation, which selects the degree of desaturation.

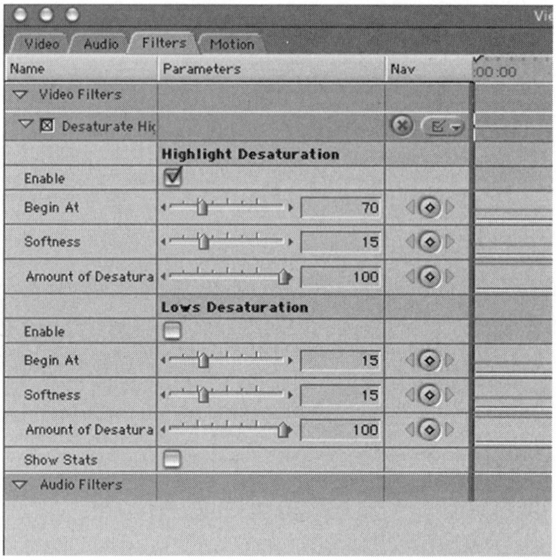

Image 4-13 *The Desaturate Lowlights and Highlights filter display.*

Broadcast Safe

This is an extremely useful filter for use when dealing with colors that may have illegal chrominance or luminance levels. The control window contains variable luminance- and chrominance-limiting controls as outlined below:

- *Clamp Above*. This selects the maximum percentage of luminance or chrominance, depending whether the luminance-limiting or saturation-limiting aspect is being adjusted.
- *Max Output*. This ensures that no values are clipped by compressing the settings between the Threshold and Clamp Above adjustment controls.
- *Start (Threshold)*. This defines the point at which either the saturation or luminance information will start to be compressed, depending on whether the Luminance-Limiting or Saturation-Limiting control is being adjusted.
- *Reduce Chroma/Luma*. This adjustment control will produce a specified reduction in chroma.

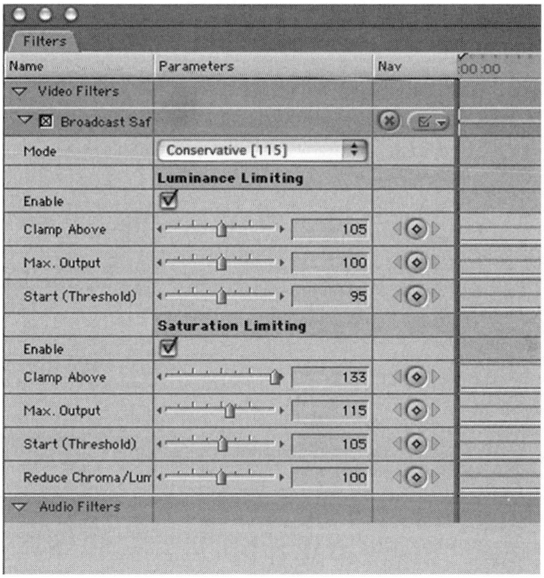

Image 4-14 *The Broadcast Safe filter display.*

Copy Filter Control

The copy filter controls are extremely useful because they allow the operator to move and copy filter settings to other clips. This type of control allows the operator to quickly move color corrections from one clip to another with just a click of the mouse or by using the Drag Filter control to move color correction settings to another clip. Here is a small breakdown of the functions included within this window.

Copy From 2nd Clip Back

This copies the color-correction filter setting from two clips prior to your current scene. This is extremely useful, for example, when color correcting a scene that cuts back and forth between two people, because the same color is generally present in every other cut. This tool allows the colorist to automatically pick up the color correction from the selected cut and copy it into the current scene being adjusted.

Copy From 1st Clip Back

This will automatically copy the color-correction setting from the previous scene or clip into the current scene or clip.

Drag Filter

When you click on this button, you may drag the color-correction setting to another scene or clip.

Copy To 1st Clip Forward

This control will copy the color-correction settings to the next scene or clip in your material. This can very useful for making an after-the-fact change in color correction when you decide that your current color correction will also work well within the next clip.

Copy To 2nd Clip Forward

A simple mouse click will automatically copy the current scene's color-correction settings to the second clip or scene forward. As with the Copy From 2nd Clip Back control, this is extremely useful when working on a scene that cuts back and forth between two people, because the same color is generally present in every other cut. This control allows the colorist to automatically pick up the color correction from the selected cut and copy it to the second cut forward.

Avid Technology, Inc.

Avid Technology, Inc. produce a wide range of nonlinear editing systems. At the top of their product line are Avid Symphony™ and Avid®|DS. Both Avid Symphony and Avid|DS have exceptionally powerful color-correction tools within the systems. Many of the color correction tools contained within Avid|DS and Avid Symphony have been implemented into Avid XpressDV 3.5.

Avid|DS

Avid|DS Version 6 provides the operator with set of color-correction tools that are powerful enough to handle even the most difficult color-correction situations. Below is a description of the various color-correction tools contained within Avid|DS systems.

Color Correction

The Color Correction effect is essentially a primary color corrector that allows you to adjust the color balance of the shadows, mid-tones, and highlights by simply dragging the color selector with your mouse to the new value, as seen in color balance tab section of the window. Also under this window tab is a luminance control that allows adjustment of the overall luminance value. The user can also adjust the overall hue, saturation, contrast, and brightness of the image with the adjust tab. Additionally, there is also a modifiable curve window located under the Curves tab in this window that allows adjustment of the red, green, and blue curves individually or together. This control also provides the ability to adjust the luminance and alpha curves separately. Finally, you have the ability to adjust the curve for the Range, with separate adjustments for the shadows, mid-tones, and highlights. Under the Masking tab, you have the opportunity to select the channel where the effect is to be applied. A mask can also be generated from the current clip's alpha channel or from another clip's external matte.

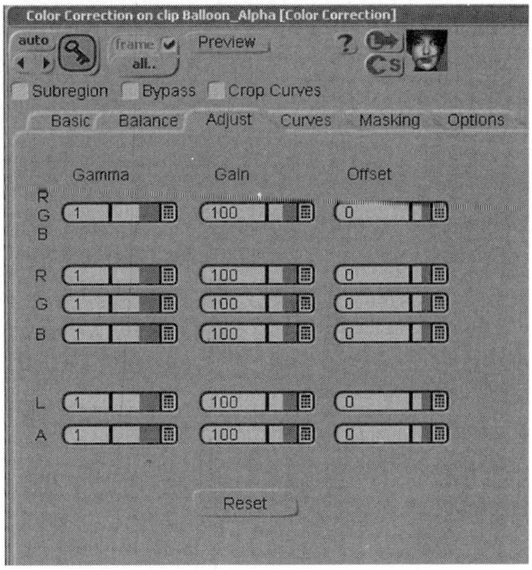

Image 4-15 *The color-correction adjustment window within Avid/DS. (Courtesy of Avid Technology, Inc.)*

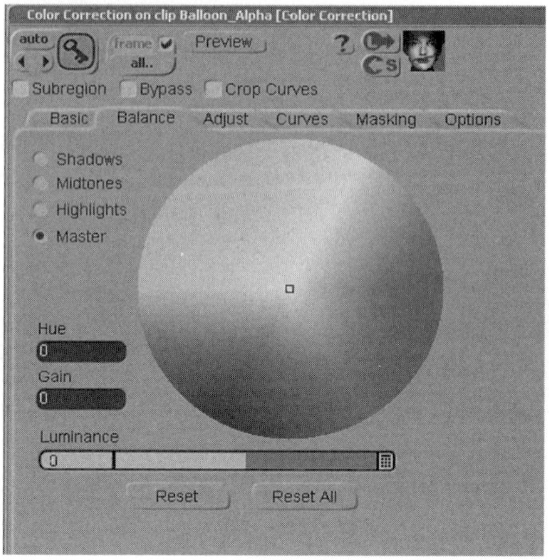

Image 4-16 *The Avid/DS color-correction color balance control window. (Courtesy of Avid Technology, Inc.)*

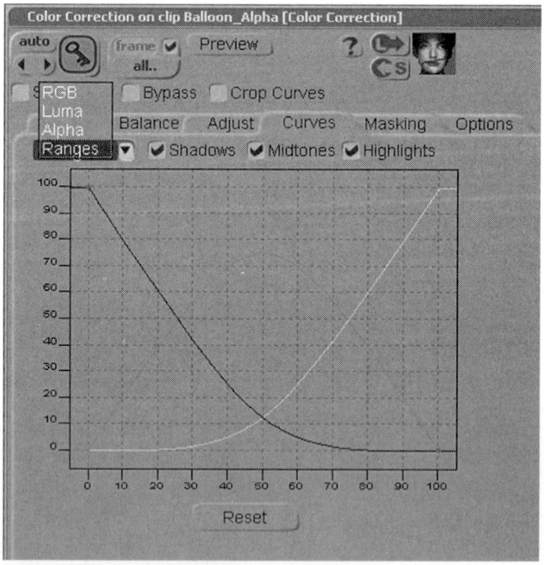

Image 4-17 *The Curve display within the Avid/DS color-correction control window. (Courtesy of Avid Technology, Inc.)*

Selective Color Correction

The Selective Color-Correction effect first shows the user a presect effects tree, which contains an input and output module. In between these two modules is a chroma keyer module, along with a color corrector module. Essentially, the user isolates the color to be corrected by using the chroma keyer to create a matte signal, and then applies the correction to it by using the color corrector.

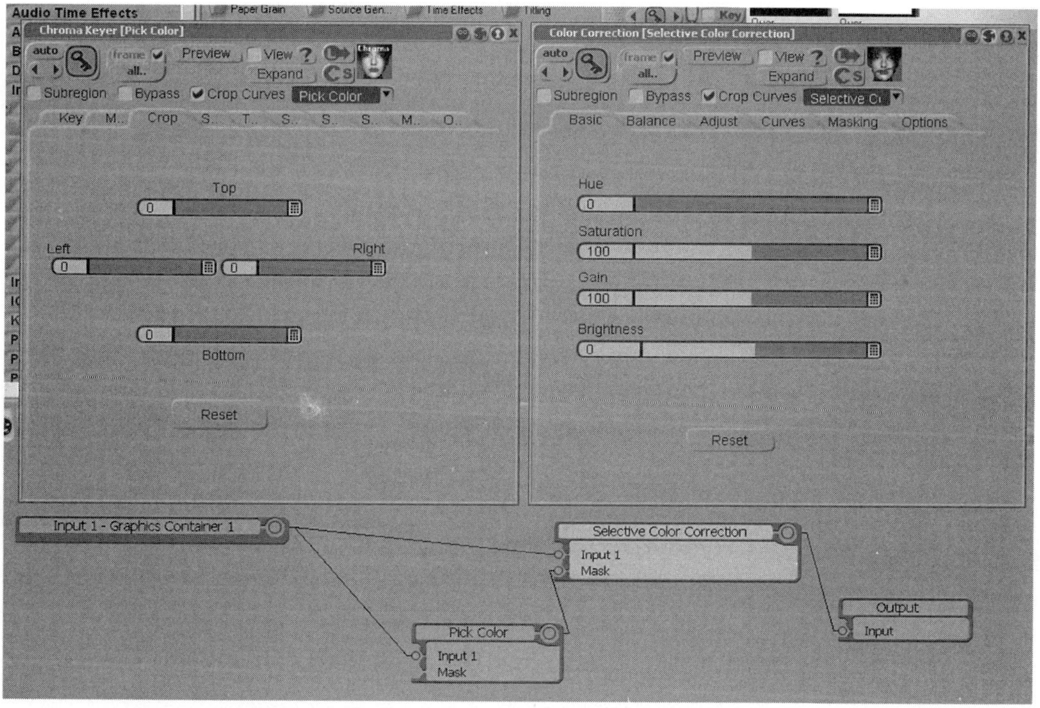

Image 4-18 *The Selective Color-Correction effect window within Avid/DS. (Courtesy of Avid Technology, Inc.)*

Complementary Color Correction

An extremely useful effect that generates a preset effects tree, providing the user with a visual guide to the color-correction path. Essentially the user creates a matte signal of the color to be changed. Following this, the first color corrector in the tree can then be used to change the color of the isolated color. The second color corrector module can then be used to color correct the surrounding information.

Monochromes

The Monochrome color-correction effect is a series of preset black and white washes that can be applied to an image. This filter can be particularly useful when you are adding sepia tone washes or other tints to an image.

Avid Symphony

Avid Symphony is the premiere product in Avid's line of nonlinear applications for real-time standard definition finishing. Avid Symphony contains an extremely sophisticated array of color-correction tools designed to easily handle even the most meticulous color-correction scenario. Among the tools is the Color Match control that allows the operator to select the color to be matched as the reference image and then apply it to the current scene's color to be changed. Using this method, the other colors are changed proportionately in order to ensure a natural correction. Within the Color Match control is also the NaturalMatch™ control, which works basically the same way as the Color Match control, with the exception that it replaces the hue values while retaining the image's saturation and luminance values. Following is an explanation of the various color-correction groups contained within Avid Symphony.

The HSL Group

This control group allows user to perform corrections to the images hue, saturation, luminance, and gamma properties. This particular color-correction group can be especially useful in situations where the image to be corrected contains color contamination such as color in the blacks or whites. It also provides the ability to perform corrections to the image's highlights, shadows, and midtone areas.

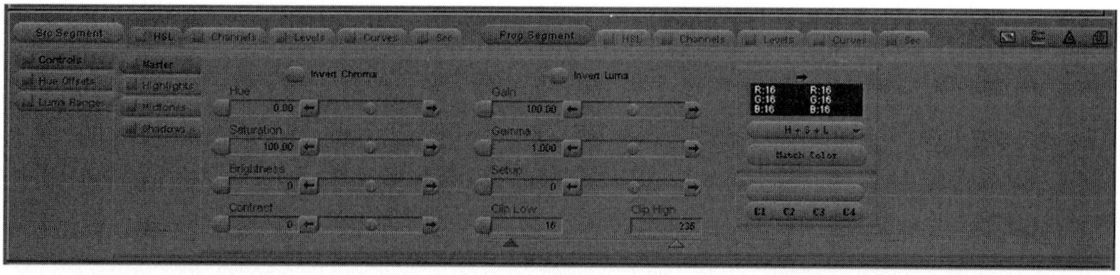

Image 4-19 *The HSL Color-Correction control window within Avid Symphony. (Courtesy of Avid Technology, Inc.)*

The Channels Group

The Channels Group controls allows the user to change the red, green, and blue output colors by blending the various components of the input channels. The slider controls within the Channels Group display provide the user with an extremely precise way of creating a blend of information for the output channel. The Channels Group also contains an extremely handy preview mode that allows you to check in monochrome for any noise that might be present in a channel.

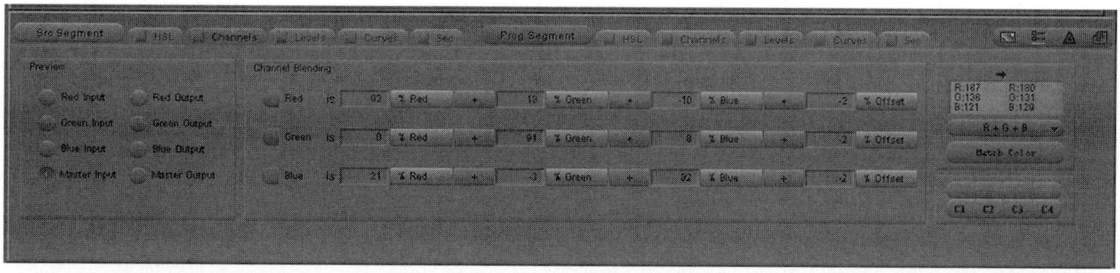

Image 4-20 *Avid Symphony's Channels Group adjustment display. (Courtesy of Avid Technology, Inc.)*

The Levels Group

In the Levels Group control area, the user has the ability to adjust the image's bright and dark areas. This can be done in the red, green, and blue channels independently or with the channels combined. The control window provides a histogram for adjusting your input and output levels. In the middle of the control window is a curve display, which has three control points for black, white, and gray. The curve window also displays the difference between the input information and the output information.

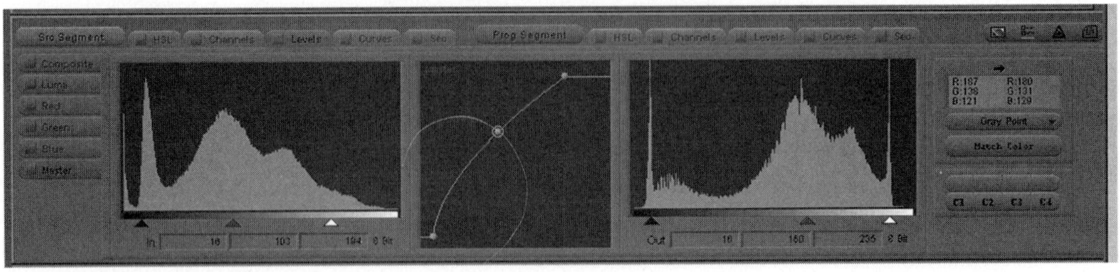

Image 4-21 *The Levels Display within Avid Symphony. (Courtesy of Avid Technology, Inc.)*

The Curves Group

The Curves Group contains a series of curve displays for master, red, green, and blue adjustment. Each curve display can contain up to sixteen control points, allowing for extremely precise color correction. Numerical values may also be added by typing numbers into the text boxes. Alternately, if the Color Match function is used, the curves will automatically be adjusted, as reflected in the input and output colors.

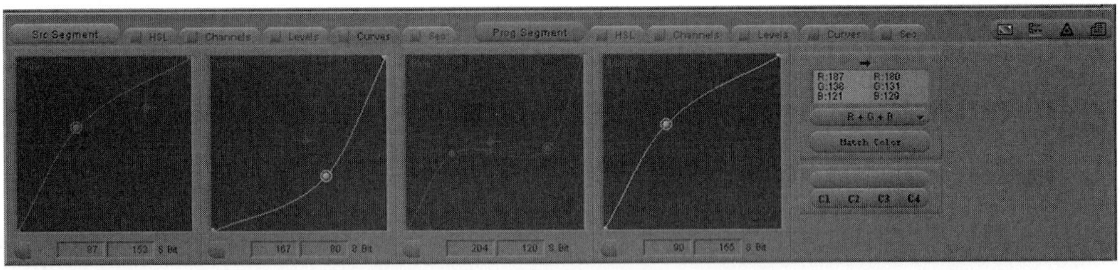

Image 4-22 *Avid Symphony's Curves Group display. (Courtesy of Avid Technology, Inc.)*

The Secondary Group

The Secondary Group is a secondary color corrector that allows color isolation to be based on an object's hue and saturation. Once the color isolation has been carried out, the properties of this color may be changed without altering the image's overall color balance. The group contains six standard vectors, which pertain to red, green, blue, yellow, cyan, and magenta. In addition to these, there are also six custom vectors, which can be customized in order to accommodate special isolation situations. The actual method of operation for the Secondary Group is as follows: The operator will first define the color range to be isolated by utilizing the Input Vector control. Following this, the operator then moves to the Output Vector controls to adjust the hue and saturation levels of the isolated color.

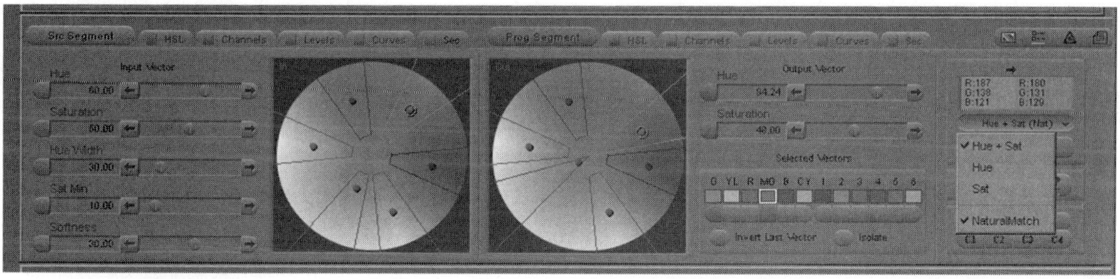

Image 4-23 *The Secondary Group color-correction display within Avid Symphony. (Courtesy of Avid Technology, Inc.)*

Plug-In Applications

Plug-in applications are essentially software applications that add to the functionality of an existing software program. In the case of color correction, plug-in applications are sometimes added to an application when there are no color-correction tools or simply to extend the color-correction power of the particular application. Below are descriptions of several color-correction plug-in applications from various manufacturers, along with a description of the tools contained within each of the applications.

3Prong

The 3Prong company makes a wide variety of AVX plug-in applications for Avid Technology systems. Among their AVX plug-in applications is the ColorFix™ color-correction plug-in, which can be added to many of the Avid Technology systems.

Image 4-24 *The ColorFix control window. (Courtesy of 3Prong.com, Inc.)*

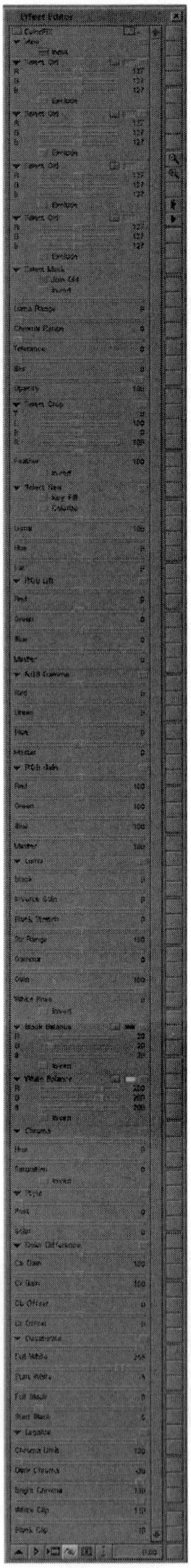

ColorFix

This plug-in application can be used with systems such as Avid Xpress DV™ for example. It provides primary and secondary color correction, along with a legalizer in order to ensure that the final correction will be within legal, or transmittable, limits. There is also an automatic black and white balance control. Simply point to the color contaminated area, and it will be automatically balanced. There is also a black stretch and knee control contained within the plug-in.

Digital Film Tools

Digital Film Tools, LLC is a division of Digital Film Works, which is a film and television visual effects company based in the United States that produces a wide range of plug-in applications for Avid®, Adobe®, Discreet® and Pinnacle® systems.

55mm

The 55 mm plug-in application from Digital Film Tools provides a large array of tools for the user. This plug-in is available for Avid products such as Media Composer™, Xpress DV, DS™ and Symphony. It's also available for Adobe After Effects, Discreet Combustion®, Pinnacle Commotion™, and Final Cut Pro. Below is a listing of the various tools contained within 55 mm. These effects have a wide range of applications and many may be particularly useful when combined with a vignette.

Color Grad

An extremely useful tool for quickly and easily generating a gradation of an image. All parameters of the gradation can be easily changed. Additionally, a mouse click on the color patch allows the operator to easily change the color of the gradation.

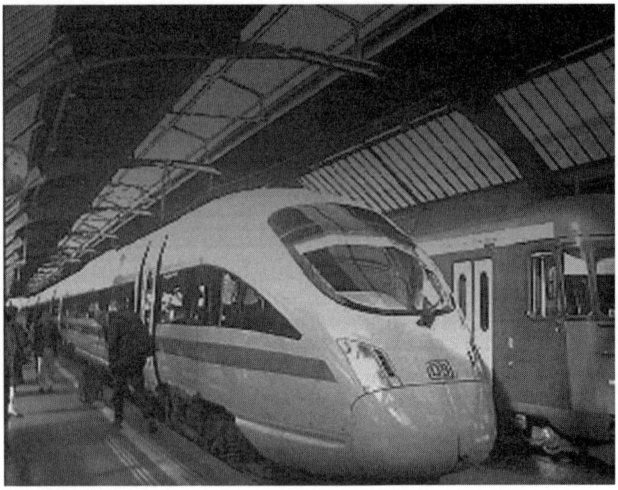

Image 4-25 *A Gradation created with the Color Grad tool within the 55 mm plug-in.*

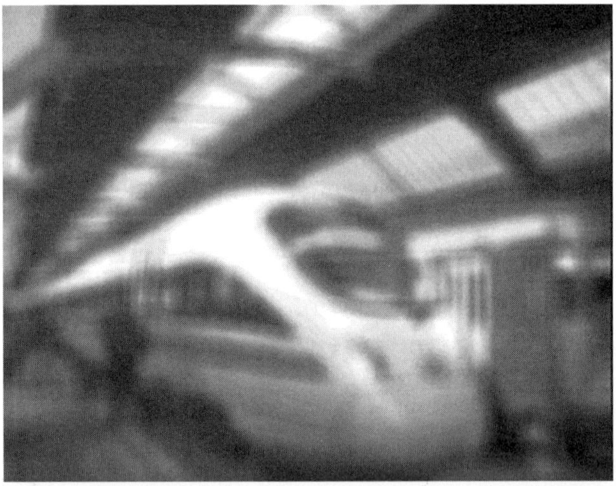

Image 4-26 *55 mm's defocus tool applied to the image.*

Defocus

This is a fully functional defocus tool, which allows the operator to customize defocus effects that can be added to the image.

Faux Film

The Faux Film effect essentially creates a fully customizable filmlike look, allowing the user access to variable amounts and size of film grain, blooming highlights, and so on. It's an extremely useful tool, especially when dealing with video-generated material that is being combined with film-generated material.

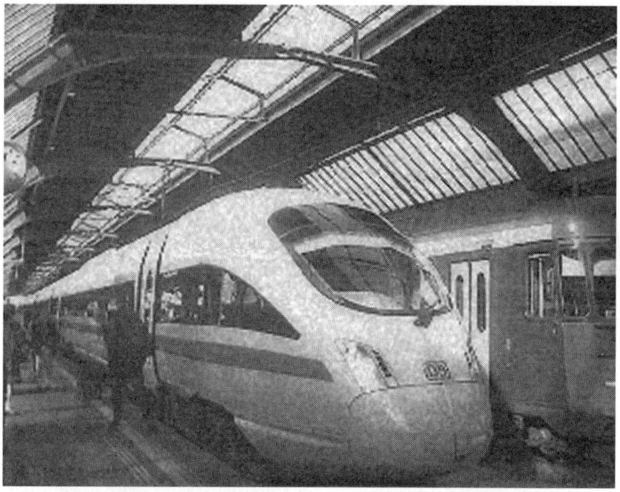

Image 4-27 *The Faux Film effect applied to an image, creating a film effect, complete with grain.*

Fluorescent

The Fluorescent effect is a rapid way to subtract the fluorescent light color within a picture. For instance, the greenish effect that occurs when photographing under fluorescent lighting can be instantly eliminated with a touch of a control.

Fog

The Fog tool creates the same effect as a fog filter placed over a camera lens. The effect can be completely customized in order to best accommodate the picture's content.

Infra-Red

The Infra-Red effect simulates the view from an infrared display and, as with all the 55 mm effects, is fully customizable.

Mist

As with the Fog effect, the Mist effect creates the look of a mist filter placed over a camera lens. This effect, as with all the 55 mm effects, can be customized to accommodate the specific situation.

ND-Grad

The ND-Grad quickly and easily generates a fully customizable neutral-density gradation that is extremely useful on many occasions.

Night Vision

This effect simulates a view through a night vision scope. It defaults to the standard green night vision display and allows for blooming the highlights and even adding varying amounts of grain to the image.

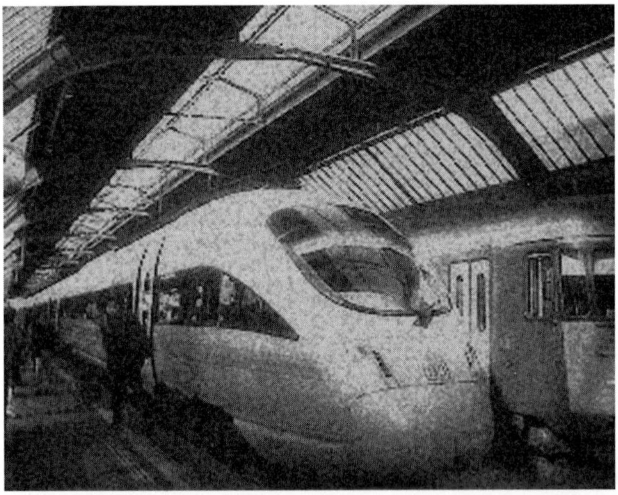

Image 4-28 *55 mm's Night Vision effect applied to an image. (See also Color Plate 20.)*

Selective Soft Focus

The Selective Soft Focus filter allows the operator to create a matte signal and then apply varying amounts of defocus based upon the area qualified. In addition to defocusing the defined area defined, the operator can also defocus the matte signal. This can be particularly handy when dealing with material that contains compression artifacts.

Skin Smoother

This particular effect can be extremely useful especially when applied to fashion material because it allows the operator to smooth out flesh tones within an image. As with all the 55 mm tools, it's fully customizable.

Tint

An extremely useful tool for applying washes such as a sepia tone. This particular effect can be very useful when combined with a vignette. For instance, the subject in the center of the image could be in color and surrounded by a sepia-colored vignette.

Warm/Cool

When this tool is applied to the image, the color temperature can easily be changed from warm to cool simply by changing one slider control. This tool is especially effective when the operator would like to create the feeling of a sunny day from what was originally an overcast day.

Composite Suite Color Correction

The Composite Suite color-correction plug-in from Digital Film Tools offers an enhanced primary and secondary color corrector for Avid products such as Media Composer, Xpress DV, DS, and Symphony as well as Adobe After Effects. The Composite Suite Color Correct component of the plug-in is a powerful primary color corrector, while the Composite Suite Selective Color Correct section is an extremely versatile selective color corrector. The Selective Color Correct component allows the operator to select the information to be corrected by defining a matte and then applying the color correction only in this area. Additionally, the matte created may be defocused to create a specific effect or to help deal with video material that may have compression artifacts.

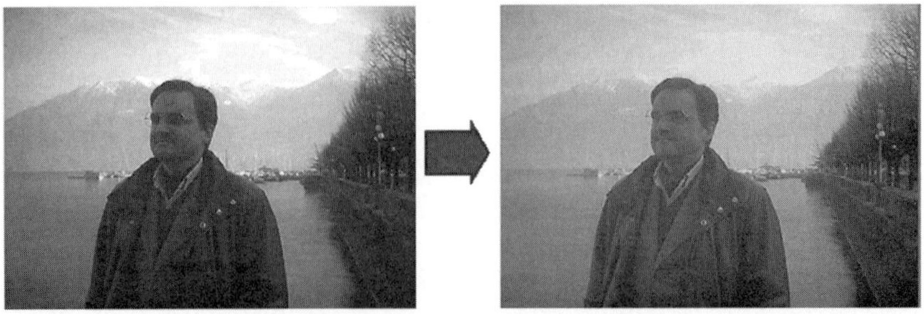

Image 4-29 *An example of the Composite Suite color-correction plug-in's primary color corrector. (See also Color Plate 5.)*

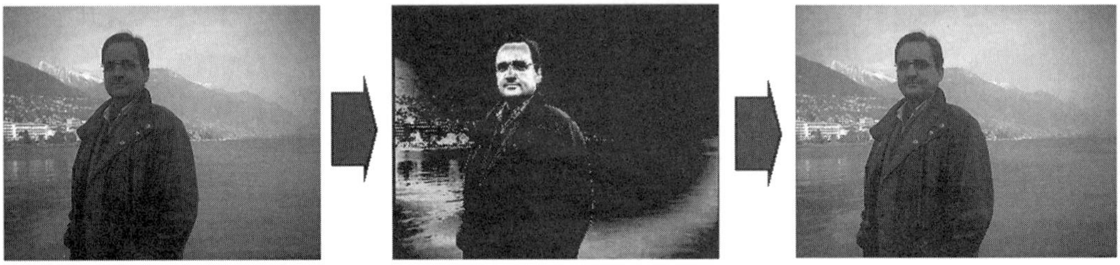

Image 4-30 *A correction made with the Composite Suite color-correction plug-in's selective color corrector.*

In-Sync Corporation: ViXen

The In-Sync™ Corporation is the worldwide distributor for the Xentrik Software plug-in application ViXen. The In-Sync Corporation has a wide variety of software products for nonlinear editing and color-correction applications. The ViXen plug-in application can be utilized with such systems as Speed Razor™, Adobe Premiere, Adobe After Effects, Incite, Silver®, Purple®, Media Studio®, Photoshop, and Video Action.

ViXen provides the operator with an easy-to-use control window, complete with a preview display and a set of video scopes. The main display has a series of tabs that guides the operator through the application. The first tab is Levels, which allows the operator the ability to adjust the material's overall video levels such as lift, gamma, gain, saturation, knee, and slope.

The Color tab allows the operator to adjust the image's color balance using the red, green, and blue slider controls. Selecting the Sharpness tab gives the operator control over image sharpness and also the ability to perform a softening of the image. Finally, we come to the Temporal and Spatial tabs. The Temporal tab allows adjustment of temporal noise reduction within ViXen, while the Spatial tab controls the spatial noise reduction.

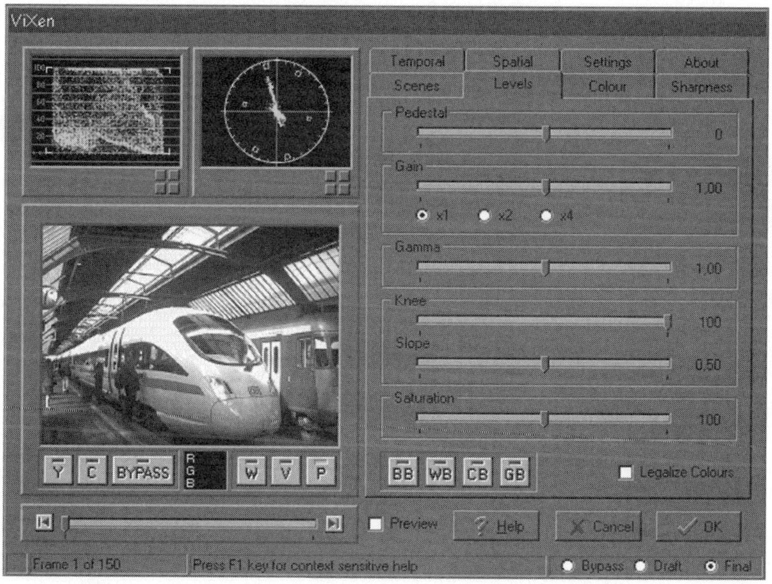

Image 4-31 *The main ViXen control window.*

Synthetic Aperture

The Synthetic Aperture™ company has an impressive array of plug-in applications that can be used with Adobe Premiere, Adobe After Effects, and Final Cut Pro. Among their array applications are Color Finesse™ and Video Finesse™.

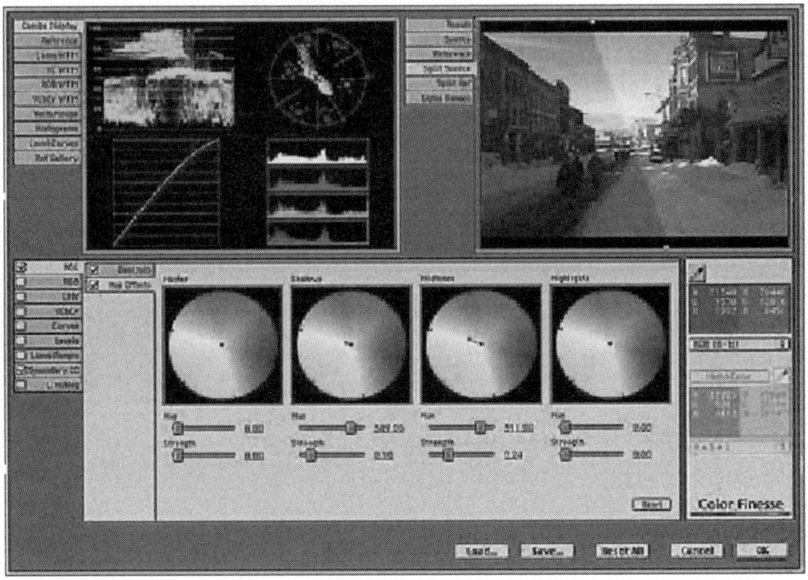

Image 4-32 *The main Color Finesse control display. (Courtesy of Synthetic Aperture)*

Color Finesse

The Color Finesse plug-in can be utilized with the Macintosh™ version of Adobe After Effects or Final Cut Pro. This application contains a comprehensive primary color corrector along with a very powerful secondary color corrector controlled through a graphical user interface (GUI). Color Finesse also includes a very nice wipe feature that allows the user to compare before and after color correction, along with a reference gallery to permit color comparison to other scenes in the material that is being corrected. Also included within the Color Finesse display are a full set of video scopes, a histogram, and a level curve that allow the user to see the fine details of the image being corrected. Color Finesse also has an extremely useful Color Match feature, which can be used for matching the color on consecutive shots. The Color Finesse color-correction modes are separated into several different areas. The user starts with the primary HSL (hue, saturation, and luminance) color correction. The HSL color-correction section is ideal for basic primary color correction and is easily adjusted using the slider controls. In addition, the tonal ranges can be easily defined, allowing for corrections only in a specific area. Following this, the operator uses the Hue Offsets control, which provides a color wheel display for adjustment of shadows, mid-tones, highlights, and master hue offsets.

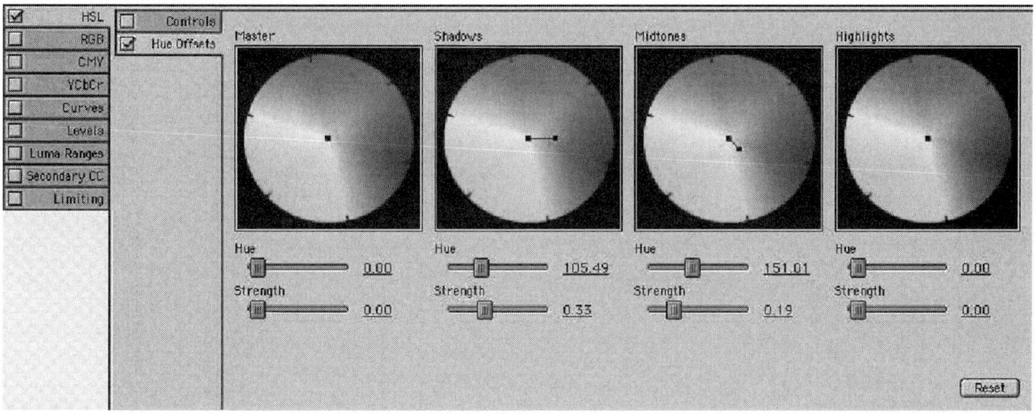

Image 4-33 The Hue Offset control display within Color Finesse. (Courtesy of Synthetic Aperture)

Next is the Channels Control section, which allows for primary color correction using individual sliders. The Curves Control has a separate curve display for the master, red, green, and blue channels. Using the sixteen definable points, very precise color adjustments can be easily made. The Levels Control provides a histogram in order to adjust the black, white, and gray areas of the image. Next is the Secondary Color Corrector, which allows for selecting and adjusting the levels of a particular color within an image without disturbing the scene's overall color balance. Finally, the Limiter ensures that any levels are brought within legal, or transmittable, limits.

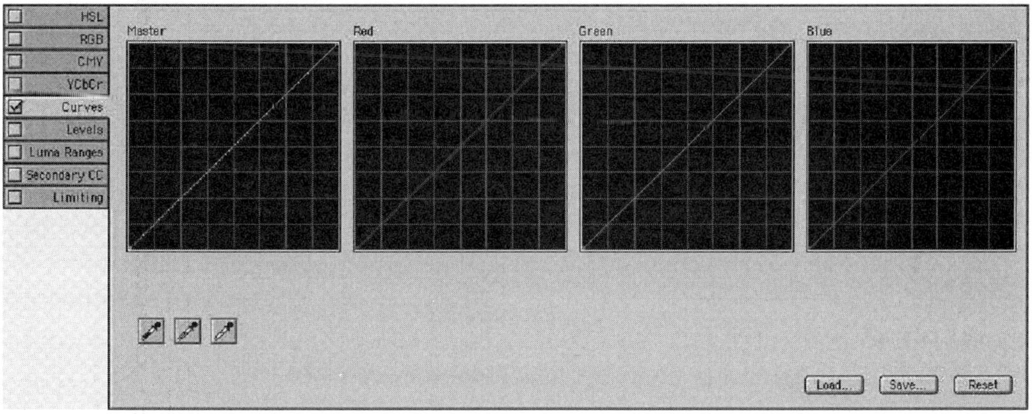

Image 4-34 The Curves control within Color Finesse. (Courtesy of Synthetic Aperture)

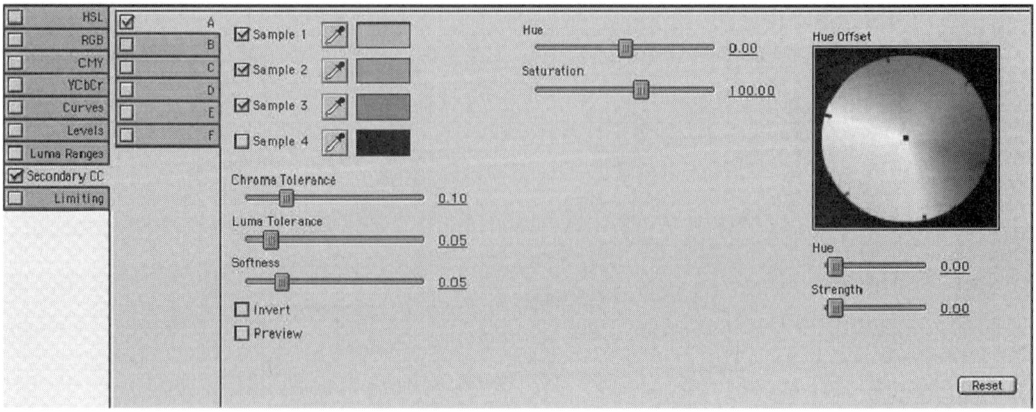

Image 4-35 *The Secondary Color Corrector window within Color Finesse. (Courtesy of Synthetic Aperture)*

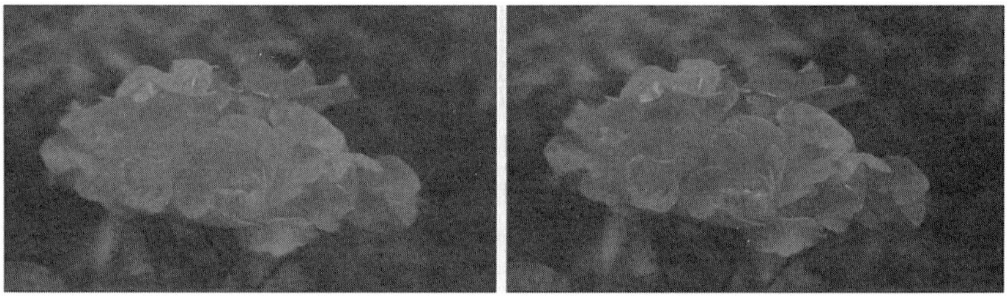

Image 4-36 *A Secondary Color Correction achieved with Color Finesse. (See also Color Plate 8.) (Courtesy of Synthetic Aperture)*

Video Finesse

The Video Finesse plug-in is available in both Macintosh and Windows® formats for use with Adobe Premiere. It contains a level adjustment section that allows for individual adjustment of black, gamma, gain, knee, and black stretch in the red, green, and blue channels, separately or combined. The user-friendly display contains a preview of the corrected image, along with a set of video scopes and a curve display. Video Finesse also contains a separate window for the use of its sharpening tool. All sharpening and enhancing parameters can be easily changed by using the window's slider controls. Alternately, there is also a softening control window, which be utilized for various applications such as creating a soft focus effect. The Gaussian Blur filter window is extremely effective for creating defocus effects. Video Finesse's Skin Detail filter provides the user with a control window that can limit the flesh tone detail. This particular type of filter can be extremely useful when color correcting fashion material. The Gamut Limiting filter within Video Finesse is extremely useful for ensuring that your corrections will be brought into legal limits for either PAL or NTSC. Video Finesse also has a Bandwidth Filter, which can be very useful in certain applications such as color correcting material containing cross-color artifacts. The Bandwidth filter counteracts these artifacts by rounding off the high-frequency information. The Dropout-Video Defect Removal filter very effectively conceals videotape dropouts with separate adjustments for black and white sensitivity because dropouts may be either black or white. Video Finesse also contains a Video Noise Reduction filter, which can be extremely useful when working with poor-quality material. It's a very simple filter to operate, with two sliders, one for edge sensitivity and another for the level of noise reduction to be applied.

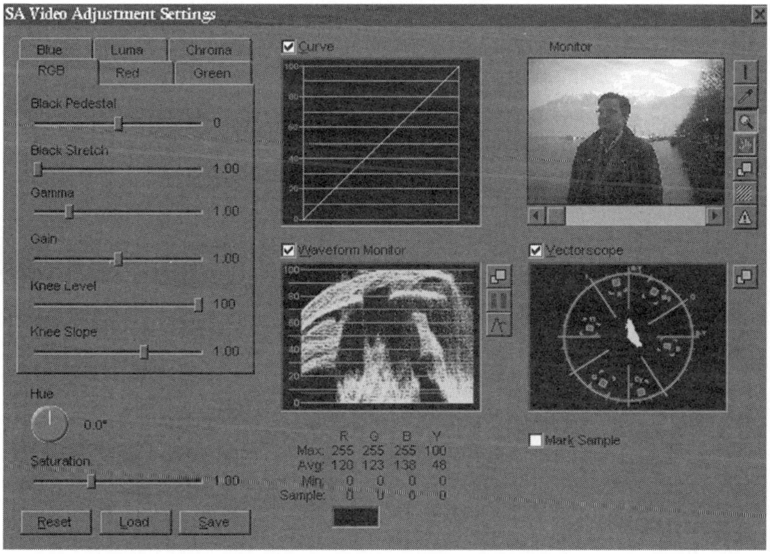

Image 4-37 *The main control window of Video Finesse.*

CHAPTER 5

Types of Color Correction

During the course of time, a colorist will find that just about every session will require a different combination of color-correction tools in order to tackle the job. A frequent problem that comes up during lengthy color-correction sessions is that after a while nothing looks right. In most cases, this is simply due to stress built up during the course of a session. A good cure for this is to just take a short break. However, while I was learning to be a colorist, my teacher always said, "When in doubt, record it! Chances are, it will look much better tomorrow." I took his advice on many occasions, and very often it did look considerably better the following day. During my early days as a colorist, a facility manager advised me that as long as the blacks and whites are balanced, the rest can be purely subjective. I found this to be quite true given that over time you'll find that no two individuals will color correct the same scene in exactly the same way. I've actually found that one of the most difficult things to accomplish is to match a color correction that someone else has done, given that you have little or no way of ascertaining what combination of color-correction tools he or she has utilized in order to achieve the correction.

In many cases, a simple change to the basic video levels can produce the desired results.

For instance, if the color of the image is basically correct but the flesh tones tend to be slightly magenta, a small adjustment to the hue control can sometimes save the day. However, in most cases, primary color correction must be used in order to balance the color throughout the material. In many situations, if primary color correction is carried to an extreme, certain critical colors in the scene may be changed. In order to correct this, secondary color correction may need to be applied. The following sections explain the various types of color correction, their uses, and when each applies.

Basic Video Levels

The basic video levels consist of hue, saturation, lift, and gain. As basic as these controls for these levels are, their adjustment can often have a dramatic effect on an image.

The Hue Control

The hue control is used for the adjustment of the chroma phase and is also referred to as "tint." The control's function is quite similar to the hue control that is found on every NTSC television. As mentioned earlier, subtle adjustments can be extremely useful for situations in which, for example, flesh tones appear to be slightly off-color.

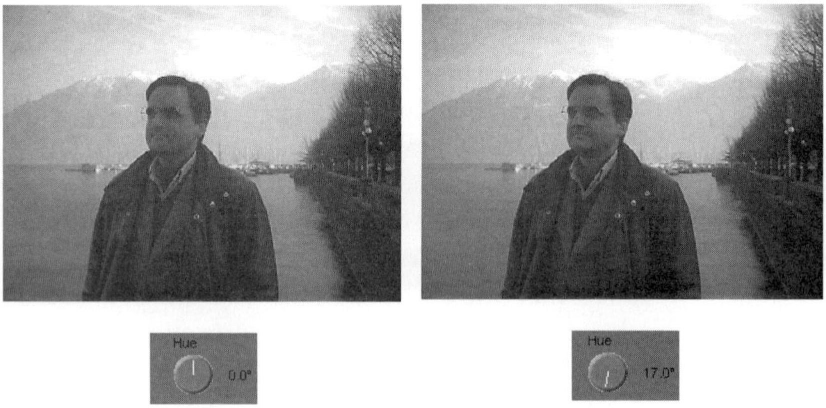

Image 5-1 *An example of a hue control adjustment.*

The Saturation Control

An adjustment to the saturation control changes the level of chroma in the image. A subtle adjustment of the saturation control can be done to balance a particular shot that may have either more or less chroma information than the surrounding scenes. A more aggressive approach is to use the control to create a special look for an image. It's important to note that care should be exercised when adjusting the saturation control because high levels can result in "illegal" chroma values that exceed the maximum signal amplitude that may be transmitted. On the other hand, low chroma levels can cause problems when secondary color correction needs to be applied. A frequent misuse for the saturation control is to excessively drop the level of chroma throughout the material in order to minimize the change of color from scene to scene. If this is done, the result is usually a very dull and desaturated picture, and this approach is not generally recommended. A useful application of this control is to slightly increase the overall chroma level throughout the picture or scene in order to aid in color separation when secondary color correction will be applied. Another useful application is to drop the overall chroma level and then apply a color wash downstream, such as a sepia tone effect, in order to create a special look to the image.

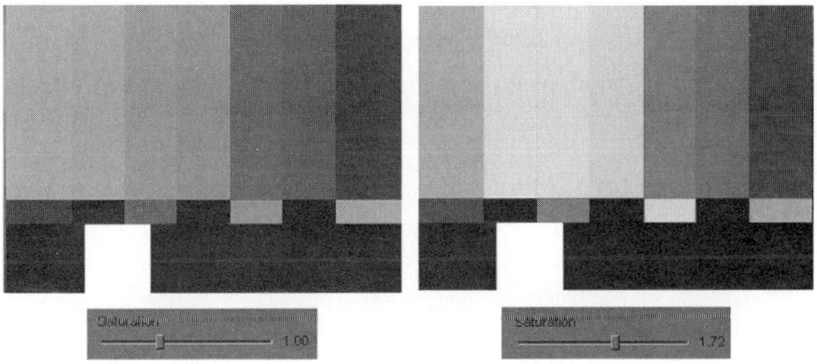

Image 5-2 *An example of a saturation adjustment.*

The Lift Control

The lift control, or pedestal control as it's sometimes referred to, adjusts the overall level of the black information within the picture. The black information should be at 7.5 IRE in NTSC and at 0 IRE in PAL, when viewed on a waveform monitor. When adjusting this control, take care that the level does not fall below these parameters. However, adjusting the control to levels above these parameters can result in a muddy looking image.

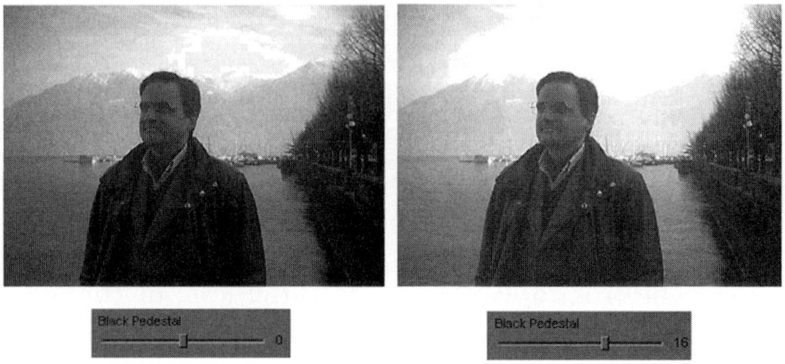

Image 5-3 *An example of a lift, or pedestal, adjustment.*

The Gain Control

Adjusting the gain, or luminance, control will change the level of luminance in a picture. Selective use of the gain control can be extremely useful for brightening up a dark image. As with the lift control, caution should be exercised to ensure that the levels fall within the parameters of the broadcasting standards and do not go beyond 100 IRE when viewed on a waveform monitor. The gain control should not be overused because excessive adjustment of a dark scene can result in the magnification of noise or artifacts that may not be normally visible.

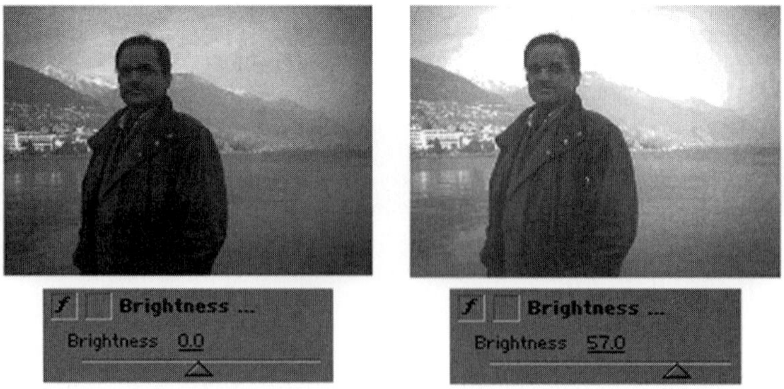

Image 5-4 *An example of luminance adjustment using the gain control.*

Primary Color Correction

Primary color correction is a method of adjusting the overall color balance within a particular scene. A primary color corrector consists of overall controls for adjusting the lift, gamma, and gain levels, plus color-balance controls for each parameter. The correct sequence for performing primary color correction is to first adjust the lift control and then balance the color of the lift, or black, information in order to ensure neutral blacks. Following this, move to the gain control and adjust the level of gain, or luminance, in the picture, then adjust the color balance of the gain information to ensure that there is no color contamination in the white areas of the picture. Finally, move to the gamma control and adjust the desired level of gamma for the picture. Using the red, green, blue color-balancing control for the gamma, you can now create the desired look or overall feeling for the image. It should be noted that there is always a certain amount of interaction between the adjustments of the lift, gamma, and gain color-balancing controls. Upon completing the gamma correction, it may be necessary to perform a slight adjustment to either the lift or gain color balance.

Of course, there are many applications for primary color correction. For example, primary color correction could be used to color correct a scene that has been shot under poor lighting conditions. Prior to correction, the scene would look quite orange in appearance. A small adjustment to the primary color balance and the image should look normal once again. Another useful application is to create a "night for day" effect for a scene that was actually shot in the late afternoon. This can easily be achieved simply by lowering the gain level and slightly dropping the level of gamma. Following this, use the red, green, and blue color-balance adjusters for the gamma control and create your blue evening feel for the image.

Image 5-5 Primary color correction.

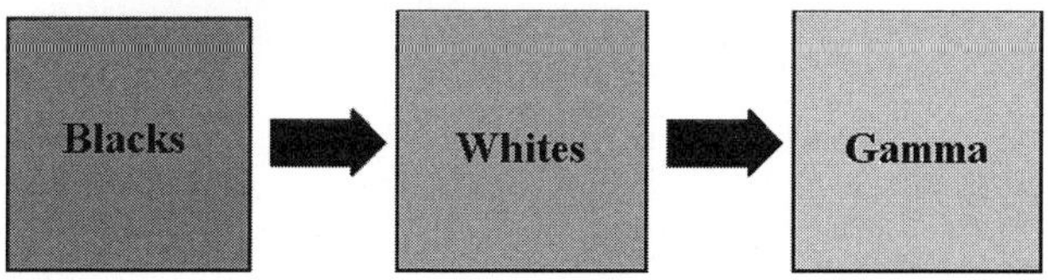

Secondary Color Correction

Secondary color correction is a method for adjusting the parameter of an individual color within an image without affecting the overall color balance of the scene. In its basic form, secondary color correction allows the operator to adjust the hue, saturation, and luminance values for either red, green, blue, yellow, cyan, or magenta. The more advanced forms of secondary color correction allow for the isolation of a particular color by its hue, saturation, or luminance differences in respect to another area in the scene that may contain a similar color. For example, you might have a still-life scene containing two apples and you would like to change the color of one of the apples, while leaving the other untouched. Upon observation you notice that one of the apples has a slightly higher color saturation than the other. Because of this difference in saturation, you can separate this object from the other. This can be seen in Image 5-6, where the person's face and coat have been isolated, or "qualified" as it's referred to in some systems.

Another way to separate colors in order to perform secondary color correction is to generate a luminance key matte signal. In this approach, the color is separated according to the luminance differences from one color to another. Once the object is isolated, you may then alter the hue, saturation, and luminance values. During the process of secondary color correction, it's important that care be exercised when working with material that may contain any sort of compression. In this situation, excessive luminance changes can result in the compression artifacts becoming magnified. The usual result is a "blocky" pattern effect, which varies in size, depending upon the amount of compression in the source material.

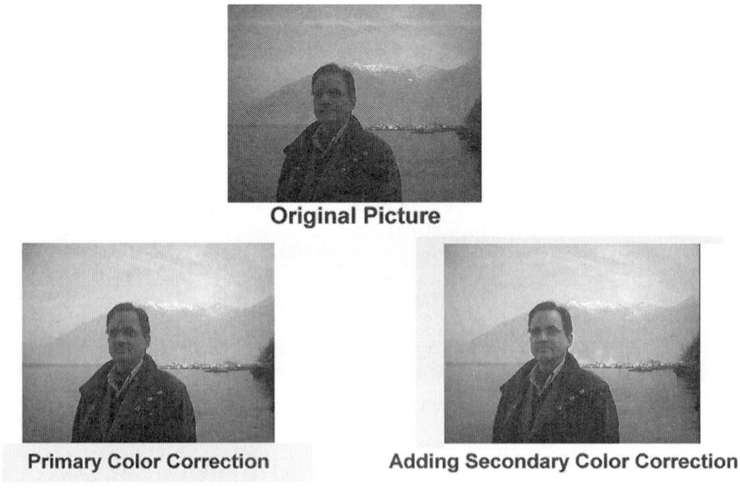

Image 5-6 *Examples of primary and secondary color correction.*

As illustrated in the chart in Image 5-7, it's important to carry out the process of primary color correction prior to starting with secondary color correction. Making an adjustment to primary color correction after starting secondary color correction can result in changes to the color isolation because the overall lift, gamma, and gain values for the scene will be altered.

As with primary color correction, secondary color correction has a multitude of applications. A very useful application for secondary color correction is to color correct a sky that is not quite as blue as it should be. The existing color information in the sky can be isolated by its hue or saturation. Following this, simply apply a small saturation adjustment. Another useful application for secondary color correction is to brighten up the flesh tones of a person who was photographed in a backlit situation. In this case, simply isolate the flesh tones using the hue, saturation, or luminance differences in the color, and then apply a small luminance correction.

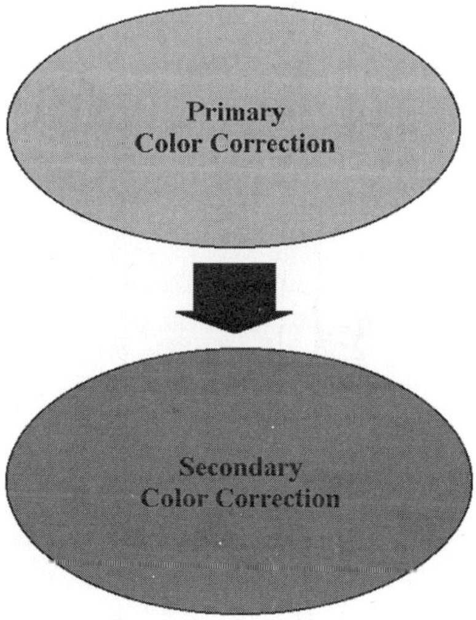

Image 5-7 *Primary and secondary color correction.*

Adjustable Curves

The curves adjustment display can be utilized to adjust an image's range of tones simply by changing the graph with the mouse. The curves graph shows the lift information at the lower-left portion of the graph and the gain, or luminance, information at the top-right portion of the graph. The information between these two areas on the graph represents the mid-tones. The adjustable curve feature also allows the user to adjust the luminance curve separately from the red, green, and blue curves, which can result in effects such as compressing the black information and stretching the white information. Keeping this in mind, extreme adjustments can be used to alter the effect to certain luminance levels. For example, this type of adjustment can be used to create a high-contrast image or to diminish a camera flare.

Adjustable curves have many specialized applications. For example, an adjustable curve can be extremely useful when color correcting material that originates from a film negative when the client would like to create a "print look." This can be achieved by applying a print-style curve to the image.

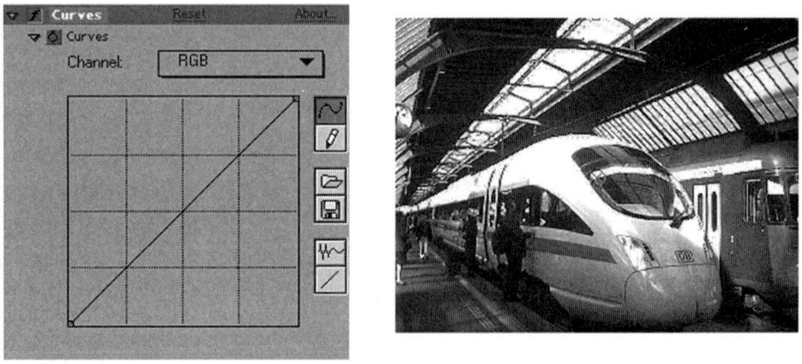

Image 5-8 *An image before using adjustable curves.*

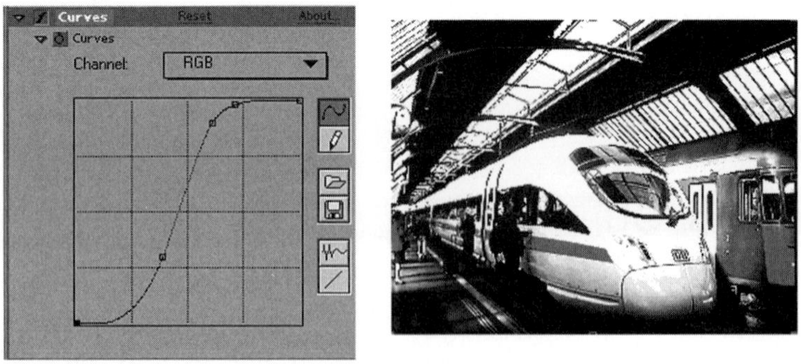

Image 5-9 *A corrected image made with adjustable curves.*

The Knee Control

When you attempt to extensively raise the level or gain of an extremely dark signal, you may magnify noise that may be hiding within the image. You can brighten the picture by adjusting the gain control followed by making an adjustment to the mid-tones with the gamma control. Very often when this is done, you end up with the highlights becoming blown out or "clipped" once they exceed 100 IRE and are cut off. In order to recover some of the detail lost in the "clipped" portion of the image, the knee control can used to recover the lost highlights. In essence it becomes a sort of a soft clipping feature and can be extremely useful when creating special looks during the color correction of fashion material.

A useful application for the knee control is to soften the highlights of video-generated material. In many cases, material shot with a video camera will have extremely bright highlights.

By utilizing the knee control, these highlights can be softened. This correction will result in a far more pleasing effect.

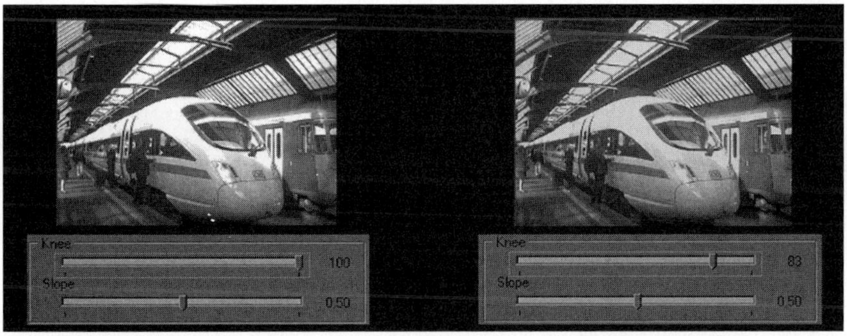

***Image 5-10** A comparison of an image with and without a knee control adjustment.*

Area Isolation

The function of area isolation is to mask off a portion of the picture in order to carry out a color correction only in this area. For instance, area isolation can be achieved by creating a luminance key matte signal. The areas defined by this key matte signal will contain the color correction. Another method of isolating an area is to manually define a mask either by using a predefined shape or by physically drawing one. When using Adobe After Effects®, an adjustment layer can be created, followed by a mask, in order to isolate an area that will contain a correction. The correction carried out in the defined area may be either a primary or a secondary color correction. An example of area isolation would be if you had an image containing two oranges that had identical values for hue, saturation, and luminance, but you wanted to change the color of only one of the oranges. Area isolation could be applied to one of the oranges and the color changed, leaving the other orange unchanged.

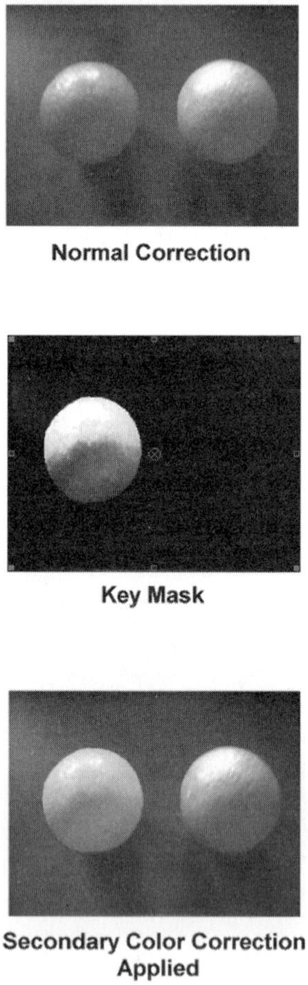

Image 5-11 *Utilizing area isolation. (See also Color Plate 3.)*

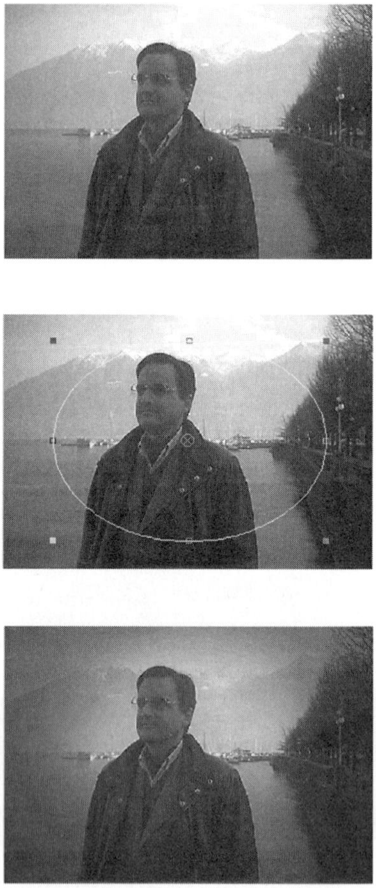

Image 5-12 *A vignette effect created by using an adjustment layer with Adobe After Effects®.*

CHAPTER 6

Seven Sample Scenarios and Operational Tips

Seven Sample Scenarios

The following scenarios were compiled in diary format in order to give you an idea of how each particular job progresses from the initial planning phase right up until its completion. While reading these scenarios, you'll see how each situation creates different obstacles and how we overcame each of them during the course of the session. In these situations, it's very important to note how closely the colorist and compositing artist or online editor work in order to resolve any problems that may arise. Although these scenarios are taking place in a dedicated color-correction suite, they demonstrate the necessity of utilizing color correction in the compositing suite prior to a tape-to-tape session, as well as the importance of the client/colorist working relationship. The majority of the scenarios pertain to commercials because they generally involve the most complicated situations; however, we've also included a documentary and a music video in order to give you an idea of how they differ.

A Beer Commercial

Monday, 9:00 AM

We called a meeting with the compositing artist and the art director from the advertising agency to discuss how this commercial should look. The art director explained how he would like a very bright yellow feeling throughout the commercial. I immediately realized that this job could present some difficulties because there were no hard cuts within the material. It would consist of a series of dissolves. This meant that it would be a challenge to maintain the consistent shade of yellow throughout the material.

Monday, 11:30 AM

The uncut material had gone into the compositing suite and was being composited. However, it had not been previously revealed to me that the dissolves consisted of a dissolve to a yellow field and then a dissolve to the next scene. Additionally, the yellow feeling for the commercial was generated by a layer being superimposed over the picture. I was a little apprehensive but decided to wait and see how the material looked once it came to me.

Monday, 3:00 PM

The material arrived and I had a look at it. I soon realized that we had a problem and confirmed it by creating a few test corrections on the material. Because of the method of layering used, it was virtually impossible to maintain a continuous level of color throughout the material. I began to rack my brain in order to find a solution that would create a pleasing result for the client. I phoned the compositing artist, and he immediately came over to the suite. I concluded that the best way to achieve the desired result was to receive all the layers for the commercial individually on videotape with matching time-code numbers. The compositing artist agreed that this would be the best course of action, although it meant more work for him. Following this, the material could then be recomposited and then a slight correction reapplied if necessary. We discussed this with the client, who also agreed, and we booked time in the suite for the following morning.

Tuesday, 9:00 AM

The compositing artist recorded each layer separately on videotape, with matching time codes. Following this, the tape was brought to us, and we commenced the color correction. Utilizing this method made the corrections quite simple, and everything went extremely fast. As it turned out, since one of the layers was a yellow field, it one didn't require any corrections. Upon completion, we played out the material to the recorder and brought the color-corrected videotape to the compositing suite. Following this, all the layers were assembled once again.

Tuesday, 11:30 AM

The client has viewed the finished material in the compositing suite and was just about satisfied with the result. A slight overall correction will need to be applied, so the material has once again been sent back to us. We viewed the material and ascertained that it was just about okay, and an overall correction providing a bit more contrast would do the trick. Once this was done, the client was shown the final result and was extremely pleased. We once again played this out, and the commercial was completed.

An Airline Commercial

Thursday, 4:00 PM

The telecine suite has received the film material for the commercial, which will be transferred with an overall color correction. They, of course, made sure that the blacks were not crushed and the whites not clipped. We also made sure that during the telecine session, the basic color correction was somewhat in the direction that the final result should be. There was also a little extra saturation applied in order to assist in any secondary color correction down the road. Actually, the project turned out to be two commercials, and they were extremely well photographed. From what I saw while I was observing the telecine transfer session, it appeared to be quite an easy job.

Monday, 9:00 AM

The film material went into the Avid suite for an offline edit, which was completed by noon. This was actually quite fast considering it was two commercials. I was told that the online session would commence after lunch.

Monday, 1:00 PM

The clients arrived for their online editing session, and they started in on the autoassembly soon afterwards. It turned out to be quite a straightforward session since there weren't any special effects or layering to deal with. They wrapped up with everything at around 2:30 in the afternoon.

Monday, 3:00 PM

The clients arrived for their color-correction session of the edited material. We viewed the first commercial, which called for additional contrast and a little more overall saturation.

We then viewed the material with this correction, and it was much better, but still a little rough around the edges. We decided that there were some shots here and there that needed some special attention. Because of this, we performed corrections on those areas. Finally, everything was looking pretty good; however, I suggested that we try to help the skin complexion of the main actress because her wrinkles were showing up a bit too much for my liking. I decided to pull off a little trick. I isolated the color of her skin with the secondaries, and then changed the contrast only in this area, which smoothed it out. This effectively concealed her wrinkles and actually gave the entire image a much smoother appearance. I showed this to the clients and they were extremely pleased, so their first commercial was completed.

We then moved onto the second spot, which was actually quite short. This one was essentially a field of flowers with an airplane flying overhead. The problem with this one was that the director wanted the flowers in the foreground to exhibit colors similar to those of a Van Gogh painting. If the internal color legalizer was switched on, these color values were impossible to create. I stopped to think about this and concluded that the best option would be to perform the color correction with the legal color limiter switched off and, when the color correction was completed, play out the material through an external legalizer. We did this successfully, and the client was extremely happy with the results of this commercial as well.

A Supermarket Commercial

Monday, 1:00 PM

I first viewed the material as it was being transferred in the telecine suite in the morning and found the content to be quite straightforward. It basically consisted of food product shots and two different scenarios that made up the story around the commercial. At 3 PM, shortly after they finished in telecine, they moved into the offline edit suite to get started with the job of cutting everything together. I heard that they weren't finished with the offline until late in the evening because there was a certain amount of disagreement regarding some aspects of the edit.

Tuesday, 9:00 AM

The clients arrived, and they soon after started with their online editing session to put everything together. This took them a little longer than expected, but they finally wrapped up at around noon. They were scheduled to come into the color-correction suite at around 1 PM for 2 hours of color correction. I thought it would be a good idea to schedule a small bumper onto the end of that time just to be on the safe side, so I asked scheduling to add a half-hour to the session, which they reluctantly did.

Tuesday, 1:00 PM

The clients arrived for their color-correction session with tape in hand. I viewed the tape, and the job looked quite easy until the client said that the flashback shots should be in black and white with a sepia-toned wash, and the transition should be a dissolve. This may not sound terribly difficult but it actually is, because doing a dissolve from full color to black and white will show all the colors slowly draining out of the picture. I had to think about a solution to this problem, and basically found that the best solution was to change the images that needed to be black and white in the online edit suite. This way, in the color-correction suite, all we would have to do was create the sepia-toned wash and, of course, correct the colors of the rest of the material. I immediately telephoned the online editor and explained the problem. Luckily, his next client was late so he was able to squeeze us in.

Tuesday, 1:30 PM

The material was fixed in the online suite and returned to us about a half an hour later. We then carried on with the corrections and displayed the color corrections for the client's approval. The client was extremely happy with the results. Following this, we played out the finished product to tape, and we ended up with another happy client.

A Car Commercial

Tuesday, 3:00 PM

We started off by having a look at what was going on in the compositing suite. I immediately realized that it was going to be quite a challenge because each scene required quite an extensive amount of correction and there were no hard cuts in the entire commercial. I started to think about how I was going to approach this. Each scene required a vignette in a different area of the image plus primary and secondary color correction. The solution was to do a small color dissolve just before the dissolve on the tape, which would dissolve to the same correction on the inside and outside of the vignette. Then, at the end of the color dissolve, I could change the position of the vignette and dissolve into the new correction at the finish of the actual dissolve on the tape. It would work out to a lot of corrections during one dissolve but would certainly do the trick.

Tuesday, 4:00 PM

The clients arrived from the compositing suite, and we got started on the grading. The clients were looking for a nice neutral but colorful feeling for the commercial, which was certainly possible with the material. The vignettes mainly needed to cover up some camera flares. I tried my approach for concealing the corrections, and it worked out just fine. I explained to the client exactly what I was doing because it was somewhat time-consuming. They appreciated being kept informed, and it was also very good for colorist/client relations because the clients realized that they were in good hands.

Tuesday, 5:30 PM

We were just about done with the corrections and thought we'd view everything and make sure it all flowed well. Everything seemed fine. However, there was one shot that needed to be fixed, and this was quickly done. Following this, we played out the corrections once again to give the clients another look, and they were extremely pleased with the final result. We then played out the material to the recorder, and the job was completed.

A Scandinavian Floor Commercial

Friday, 6:00 PM

We were told that a rush job was coming through. It turned out to be a commercial for a company that makes Scandinavian wood floors. I thought at first that this would be quite an easy assignment, but little did I know that it would turn out to be one of the most unusual assignments.

The clients turned up a little late with the tape, which had been composited at another company.

Friday, 6:30 PM

After we discussed the grading possibilities for the commercial and viewed the material, we decided to get started. Things were going along quite well; however, I noticed that the clients appeared too preoccupied with the overall color of the scene rather than concentrating on the product being sold.

Friday, 7:15 PM

We arrived at one shot that was bathed in colored lighting. I explained to the clients that it would be extremely difficult to correctly ascertain the wall color because of the lighting used. They agreed and then told me that they had actually brought the walls with them! I was a little shocked at this statement and even more shocked when the large chunks of concrete were wheeled in, leaving a trail of concrete powder behind them. I then viewed the walls under a neutral lighting source and continued with the color correction. I felt that bringing the walls along wasn't really necessary, as I was pretty close to the color the first time around.

Friday, 7:45 PM

After the scenario with the walls, things were going along quite well. I decided to throw a little insurance policy in, so I ordered dinner for the clients, which often works miracles. While they ate, I continued working. We ended up finishing with the last correction at about 8:45 PM. At this time, I played the material back for the clients to give them a last look before we recorded the material. They were extremely happy except for one shot, which we fixed and then played out to the recorder.

A Documentary

Wednesday, 9:00 AM

The client arrived with her material and we started to have a look at it. The material looked pretty good although the colors were quite flat throughout. The client said that she'd like fairly warm colors during the interviews of the various people and that the background material surrounding the interviews should be a little desaturated and contain a fair amount of contrast. We started correcting the material, and everything was going quite well, except that during one interview the color balance for the camera was a bit off. We were able to fix this with no problem whatsoever just by using primary color correction and matched it to the feeling of the other interviews. We kept on working along until noon and decided to break for lunch.

Wednesday, 1:00 PM

We picked up where we left off and continued with the corrections. We ran into a problem with one of the outdoor shots because it was so flat that even additional contrast didn't help much. I had to think about a solution for this and tried putting gradation over the top half of the picture. I thought that this looked quite nice, and the client agreed with me. Later on, during one of the interviews, we ran into a lighting problem. The face of the person being interviewed was a bit too dark, although the rest of the picture was fine. I found a quick solution for this by isolating the flesh tones with the secondary color corrector and applying a small luminance correction that brightened up the face. After about two-and-a-half hours, we decided to take a 15-minute break in order to rest our eyes and make sure that we hadn't become the victims of color adaptation, which can occur after looking at colors for an extensive period of time.

Wednesday, 3:45 PM

We continued with our color correction following the short break and started off by looking at the last bit of material that we had corrected. It looked great, so we carried on with the corrections from where we had left off. We ran into a small problem toward the end of the documentary where it dissolved from one of the interviews to the background footage. We had to hide a color dissolve within the transition, which can be a bit tricky at times, depending on how extreme the correction is. On the first try, I noticed that the corrective change was indeed noticeable, so I shortened the dissolve. I think I shortened it a bit too much because a quick color change was evident, so I lengthened it a bit, which worked out fine.

Wednesday, 5:00 PM

We were just about finished and we were working on the last correction. So far we'd actually made 1500 color corrections, and we were ready to play out to the recorder. We got our record tape all set up and sat back while recording the material. The client was extremely pleased with the results, and off went another happy customer!

A Music Video

Thursday, 9:00 AM

Our clients arrived with the tape, and we discussed what type of colors they were looking for. Following this, we got started with the corrections. I immediately saw that the material had lighting problems, and we discussed this with the clients. I had to be very diplomatic about this because the cameraman was present during the session. The clients wanted an extremely colorful feel for the music video, and some of the material was so poorly lit that the resulting image would become quite noisy. We decided to make a bit of a compromise and back off a little bit on the correction. We then balanced the other material to match the correction that we had made on the poorly lit material, which worked out quite well.

Thursday, 11:00 AM

Things were moving along very well with the corrections and the clients were extremely pleased. We got down to a completely different setup in the music video, which called for a different strategy. The clients wanted to have compressed blacks and bright whites in this part, so we used the adjustable curves, which were quite effective. I also suggested doing a light vignette around the picture during this sequence; the client really liked it, and it looked great!

Thursday, 1:00 PM

After a short break for lunch, we continued with our corrections. I always like to have lunch with the clients because it helps to break the ice, and the session usually moves along quite fast after that. Anyhow we were almost finished, and we decided to have a look at the material. We viewed everything, it looked just fine, and we've put the finishing touches on the last correction. The client was pleased with the outcome, so we received the go-ahead to put the color-corrected material on tape. We then sat back and watched the corrected material being recorded.

Hints and Tips

The following is a compilation of hints and tips that we've collected over time that can be applied to various situations. Sometimes these tips are quite basic, and other times they might be contrary to the standard way of operating but in certain applications could help you out of a difficult situation.

Selective Color-Correction Effects

It's possible to create very interesting color effects using your selective color corrector. The first step is to create matte signal and then defocus or blur the matte. When the color correction is applied, the defocused edges will appear to create a halo effect. There are numerous useful applications for this scenario. One that's very effective is to isolate a model's hair color. Very often, the color can only be partially defined, which is what we want in this case. We then blur the matte signal a little, apply a bit of correction, and the model's hair will appear to be streaked.

A Difficult Color Isolation

When performing secondary color correction, you occasionally run into a situation in which you can just about isolate the color, but there's a small area which just won't make it. A good way around this is to adjust the primary color correction slightly in order to improve the isolation. The end result may not exhibit the correct primary color correction, but the separation of the color for secondary color correction may be perfect. Following this, you can layer another primary color correction on top of the previously performed correction, and then change the overall color balance back to the desired color.

Color Isolations from Compressed Video Material

Isolating colors when working with video material containing compression can be extremely difficult. An extremely good tip for use when you are utilizing selective color correction is to first select the matte signal. You'll immediately see the compression artifacts, which show up as blocks in the picture. A small adjustment to the matte defocus will blur the matte signal slightly, effectively hiding the artifacts on the edge of the correction area.

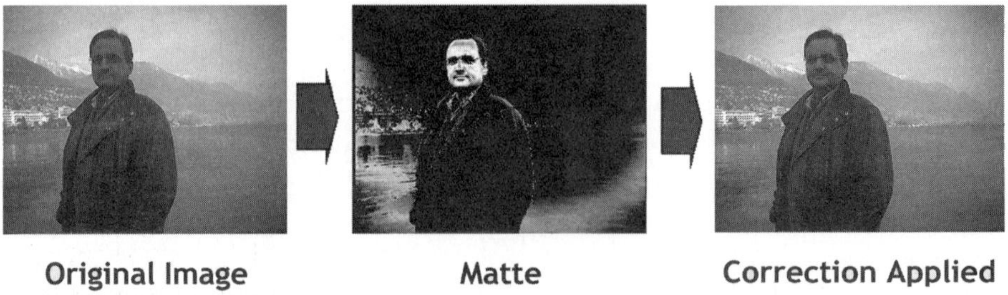

Image 6-1 Color isolation of compressed video material.

Matching Colors from a Product Provided by the Client

This scenario can frequently be a challenge, especially if the product in the video material was photographed under colored lights. However, in most cases, the lighting is usually quite neutral because, generally, it's a pack shot meant to show off the product as it's seen in the stores. When the client brings you the product model to match colors to, it's important to view the model under a lighting source balanced for daylight color temperature. Using this method for viewing the model will ensure that you're viewing the object under as neutral a lighting condition as possible.

Matching Colors Based on a Reference Tape

As easy as this sounds, it can frequently be quite challenging. The basic reason is that you have no way of knowing what combination of tools was used in order to arrive at the final correction. Also, further complications can arise if the correction was made without using any legal color limiting. In many cases, a compromise may have to be made in order to arrive at a correction that will be within legal, or transmittable, limits.

Information Backups

During the course of working, it's extremely important to create backups of your work. This way, when the unforeseen occurs, you can simply pick up where you left off. There's nothing more embarrassing than to break the news to your client that all the work that you labored over for hours has vanished with the blink of an eye. Not only is this embarrassing, but it also destroys the client's confidence in the operator. When this type of confidence no longer exists, work can be extremely tedious. If the unforeseen does occur, I always explain to the client what has occurred and recall my backup. If there is anything that needs to be done that might need some additional time, I offer some coffee to my client to help soothe his or her built-up frustration.

Setting Up Your Monitor

The color bar signal is a test signal that is used throughout the video production and postproduction industry in order to ensure that what you see is what you'll get throughout the entire process. This signal is also extremely useful for setting up your monitor. There are numerous forms of color bar test signals; however, for this purpose the most useful is SMPTE (Society of Motion Picture and Television Engineers) color bars. At the lower-right side of the SMPTE color bar signal is a series of small bars that is known as a pluge signal. This can be extremely useful for setting up the monitor brightness. Once the center bar is adjusted to the point of becoming black, proper brightness adjustment has been done. The contrast adjustment can be then be done by using the third bar in the pluge signal. Then adjust the contrast so that the third bar is just about viewable. Adjustment of the hue and saturation can be done simply by using the "blue only" button, which is found on broadcast monitors. Essentially, you should see four blue bars of identical luminance levels. If they are off-balance, simply adjust the outer bars for chroma and the inner bars for hue. Following this, turn off the "blue only" button, and your monitor should be all set up.

Image 6-2 *SMPTE Color Bars.*

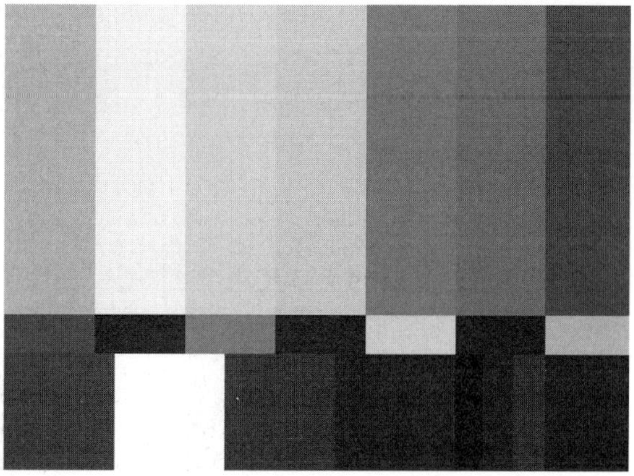

Image 6-3 *"Blue only" selected while viewing SMPTE color bars.*

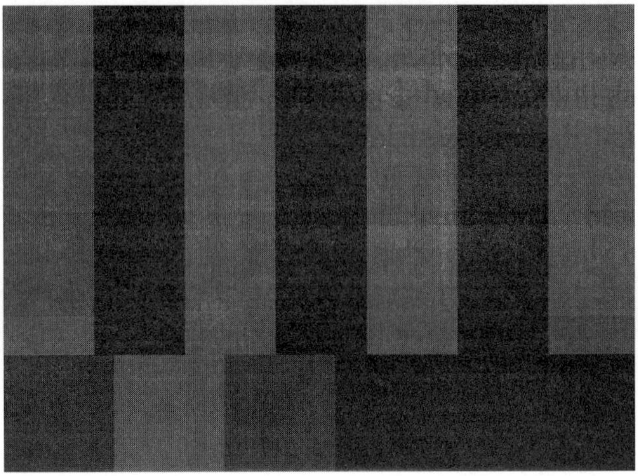

The Client

Being able to successfully work with clients is almost an art form and to thoroughly cover this would almost require a book unto itself. You'll find that dealing with clients can vary immensely because they range from those who are quite pleasant and reasonable to the difficult client who can turn your hair white after a few short hours in the color-correction suite. Within a short period of time, you'll be able to pinpoint your various clients and differentiate the difficult ones from the not so difficult ones.

There are numerous ploys you can use to successfully deal with clients. One method I have used is to put a large quantity of tabloid newspapers in the client area; these usually have headlines pertaining to subjects like celebrities being kidnapped by space aliens, and so on. The clients have really liked these, and a few have even asked to take them back to their office. This creates a light atmosphere, which ends up being good for everyone all around. I've also seen, on many occasions, that companies have installed video games in the client areas, and one facility even had a pool table! There are also some facilities that have changed the color-correction suite decor in order to provide a different theme in each suite. On one occasion, each suite was done in a different film theme, complete with some of the actual props from the various films. This type of decorating technique can help to lighten up the working atmosphere.

I've found that, when all else fails, refreshments can work miracles. A good cappuccino, a cold drink, or even a light snack can really help you to get through a session with a feisty client. I worked in one facility that had a completely stocked refrigerator, and the clients were urged to raid the icebox whenever they wanted to! There was another facility that had a restaurant-style cappuccino machine with someone manning it day and night! Some facilities, however, have decided to go all out and even put complete restaurants or coffee shops into their facilities.

Once you have the client/operator working relationship down, the next step is selling the client the color correction you've made. Very often, getting your client's approval of a color correction can end up being a real challenge. One method for achieving this that works quite well is to create three or four variations of the color correction and show them the one that you like last. They'll generally agree on the last one. Another method is to create a base correction that is a little green with milky looking blacks, and then show them your new correction. They'll generally react with horror when they see the base correction and then with delight when they see your new correction.

Working with Multiple Clients

This can also be referred to as correction by committee. This is pretty much the most stressful form of a client-supervised color-correction session. As stressful as this can generally be, it's always best to relax and take everything as it comes. One of the best ways for approaching this is to find one of the clients within the group who can act as a go-between in order to sell your correction to the rest of the group. Trying to sell your correction to each individual client within the group can at the very best be tedious and also waste valuable time. It's important to remember that if the session runs late, you're generally the one who gets the blame. A very good option when a disagreement becomes inevitable among the group is to show the clients both corrections and then offer to split the difference between the two corrections. This type of compromise generally works about 95 percent of the time. Another alternative is to create several different versions of the color-corrected material that can be put on tape, allowing them to decide later which one to use. I personally used this scenario on one occasion, and it was quite effective. Although it was time-consuming, it defused a fairly tense situation. It was interesting though because two of the clients requested a color correction that contained a fair amount of specialized corrections. When I phoned the production company the following day in order to ask how the edit session went, I casually asked which version they used. They told me that they used the normal, but colorful version that I had first tried to sell to their clients.

When to Use Which Type of Color Correction

When to use which type of color correction can sometimes be a difficult decision. For example, trying to create a blue sky when there is no color information in the sky area is not the ideal scenario for using a secondary color correction. In this scenario, a luminance key matte would be more appropriate because any degree of color could be placed within the keyed area. On the other hand, in a scenario where the person's flesh tone needs to be brightened up without affecting the rest of the scene, a secondary color correction with a small amount of luminance correction can be very effective.

The best way to decide when to use which form of color correction is to examine the contents of the image and evaluate what would be the best course of action. Of course, if there are other corrections that have been previously done on the material, these should also be taken into account because the new corrections could affect the existing ones.

Lighting Problems

Generally problems that surface during a color-correction session relate in some way or other to lighting complications during the initial shoot. Often difficult situations can arise because of mixed lighting being used. This can often be seen when you have artificial lighting inside and daylight streaming in through windows, for example. Situations like this can truly be a challenge because the effect created can be diminished by the lighting. However, in many cases, complete eradication of the effect may be impossible to achieve. Very often situations like this that occur during a shoot simply can't be helped because the lighting may be difficult to control due to the reflection, transmission, or even absorption of light waves that can affect the scene. There are numerous tools that the cinematographer can use to control a scene's lighting. For instance, a simple reflector can be indispensable for balancing the subject's lighting. On the other hand, a black card or board is sometimes used to channel or block the lighting from reaching the undesired areas of the picture.

When working with cinematographers, it's important to be familiar with the different types of lighting used. First of all, the color of light is measured by a numerical temperature scale known as Kelvin. When the light is warmer or contains more red, this will result in a lower temperature, and, on the other hand, when the light contains more blue, the temperature is higher. For example, photoflood lamps are measured at around 3200 degrees Kelvin. Alternately, daylight is measured at around 5500 degrees Kelvin. Film stocks are actually available in two forms, daylight or tungsten color balance. Cinematographers very often use tungsten color balanced film for their entire shoot, and when shooting outdoors they use an orange colored 85b filter in front of the lens in order to correct for the difference in color temperature.

The following sections describe some of the lighting terms employed by the cinematographer.

Front Lighting

This is actually the most basic type of lighting, which hits the subject head on. Although the subject may be well lit, this type of lighting is usually quite flat.

Side Lighting

Side lighting is the lighting method that hits the subject from one angle, resulting in shadows that can create a very dramatic effect. It also brings out the subject's contours, resulting in a very lifelike reproduction.

Back Lighting

This is a lighting source that emanates from behind the subject. It usually results in a very soft rendition of the subject's details because the brighter background usually tends to create a silhouette effect.

Direct Lighting

This type of lighting is usually quite hard and generally seen in electronic newsgathering applications when there is a single light source illuminating the subject.

Indirect Lighting

Indirect lighting is a diffused lighting method generally created by bouncing the light off a reflector. The result is generally a soft, low-contrast effect that is quite pleasing to the eye.

The Key Light

This is the main light on a set that illuminates the prime subject of the scene. It's the cinematographer's tool for riveting the viewer to the prime area of the image.

The Fill Light

The Fill light actually fills in the shadows created by the main key light in the scene. Using this light will result in decreasing the scene's contrast.

Frequent "Traps" Colorists Find Themselves In

Color Correcting with an Uncalibrated Monitor

Performing color correction with a monitor that is either miscalibrated or not calibrated at all can produce disastrous results. What can happen is that you will think that your corrections are looking great because they do in fact look just fine on the viewing monitor. However, once the tape is recorded and taken to the client, the true color is viewed. It happens that you are compensating for the offset in the monitor, and when the material is viewed on a calibrated monitor, you'll see the inverse effect of your compensating corrections. If you're working on a session and are in doubt about the calibration of the monitor, refer to your waveform monitor and vectorscopes. If the information on the scopes doesn't agree with what you're seeing on the monitor, there's a good chance that the monitor has a calibration problem. The best course of action is to check to make sure the monitor is properly calibrated prior to starting the job.

Overuse of the Secondary Color Corrector

This type of situation happens most frequently when one is color correcting cartoons. Occasionally, a colorist wants to correct every single color in the picture using individual vectors of secondary color correction, and after a while the vectors start overlapping. Occasionally, when this happens, noise starts to appear in certain areas of the picture. After a while, everything is overlapping so much that it becomes a total mess. Ascertaining which vector is the culprit can be very time-consuming. At this point, it's usually best to just clear out all the secondary color corrections and start over. A good rule of thumb is to do the majority of color correcting with primary color correction and then use secondary color correction for touching up the specialized areas in the image.

Color Adaptation

This can occur when one is color correcting over a long period of time. What happens in this situation is that the retinal pigments in your eye that are sensitive to light become bleached. The result is that your color perception will slowly start to drift off. The best way to counteract this is to either take a short brake or have a look at color bars on the screen to refresh your perception. This can also affect your client's perception. Sometimes the client will walk outside for a cigarette and return saying that the picture color is all wrong. This happens because the little autocorrector in the client's brain has changed the color temperature after being outside for a while and when the client returns to the suite there is a short delay before readjustment occurs. It's actually quite amazing how far off your color perception can drift when color adaptation takes effect. It's quite interesting because during the process, you feel that the color corrections are flowing beautifully, and when you later look at the material, you can't believe it.

Tips for Scene-by-Scene Color Matching

When matching the color for a series of scenes, it's, of course, extremely important that the color flows from shot to shot. What we actually mean by this is that, when viewing the material, none of the scenes should stand out as having a correction that is vastly different from that of the surrounding scenes. In order to achieve this, the operator should examine the image as a whole. This method is very important, especially when mixed lighting situations are being used within the scene. Mixed lighting can result in a bluish cast coming in from a window being combined with a warm artificial light source present inside a room. For example, we have a cut away shot of a model that may exhibit more daylight streaming in through the window than the previous shot, which exhibits more of the room's warm artificial lighting. In such situations, the entire image should be examined when matching the two scenes.

When performing color matching, the reference still store is of course an important tool for the colorist because it allows him or her to easily refer to images from any point in the material in order to ensure that the color-correction process is consistent throughout the program. Viewing the reference still store is also an extremely good way for the colorist to refresh his or her eyes in case the color adaptation has begun to take effect.

CHAPTER 7

Tutorials

The following is a collection of tutorials set up in a step-by-step fashion that allows the reader to easily reconstruct the corrections as described. Although these corrections were produced using Adobe After Effects® and Adobe Premiere®, they can be easily achieved on just about any of the other color-correction applications mentioned in this book. In addition, there are also references to the companion CD-ROM, which contains QuickTime™ movie examples and special instructional Web pages that show many of these corrections being constructed.

Primary Color Correction

This particular tutorial is uses Adobe Premiere 6.0. Although we could use the color balance controls contained within Adobe Premiere 6.0, we'll be using the plug-in from Synthetic Aperture called Video Finesse™. At the start of this tutorial, we've already imported the clip we're going to work on, and have decided that we would like to perform a bit of color correction. In addition to the steps described in this text, this tutorial can be seen as a QuickTime movie, which is located on the companion CD-ROM (see the Video Finesse Adjustment Tutorial).

Step One

Our first step is to bring up the Video Effects window in Premiere. If it's not already on the desktop, it can be found by selecting Show video effects on the Window pull-down menu on the top of the screen.

Step Two

Now, we want to select the tool we're going to use. In this case, we're going to use Video Finesse from Synthetic Aperture, located within the section called 3rd Party. We can then select the SA Adjust tool and apply it to the clip. This is done by positioning the mouse over the tool's label, holding down the left mouse button, and dragging it over to the timeline.

Step Three

Once the tool has been applied to the scene, the control window will appear. In this case, we see the Synthetic Aperture's control window complete with a waveform monitor and vectorscope. First, we'll select the RGB tab on the top of window. Next, we want to adjust the black pedestal control, which is sometimes known as the Lift control. Adjustment of the blacks to the desired level can be done by using the mouse and clicking with the left mouse button to move the slider control, while making sure the blacks are not crushed.

Step Four

Next, we use the mouse to adjust the luminance levels by clicking on the slider with the left mouse button and moving it to the desired position, while checking to make sure that the level isn't clipped.

Step Five

Now that we've got the blacks and whites set, let's set the gamma level by using the mouse in the same fashion and adjusting the gamma level so we have a nice crisp image.

Step Six

Now it's time to get the color right. On the top of the Synthetic Aperture window you'll see a series of tabs, three of them being Red, Green, and Blue. First, we'll select the Red tab and adjust the red pedestal level to ensure that there is no red in the blacks, then we'll switch over to the Green tab to ensure there's no green in the blacks, and then we'll do the same for blue. The result will be nice neutral blacks that are not crushed.

Step Seven

Next, we'll repeat the process for the red, green, and blue gain controls to ensure nice clean whites.

Step Eight

At this point, we have a nice neutral picture with clean whites and blacks. We can now concentrate on creating the desired look for the scene using the Red, Green, and Blue Gamma controls. Adjustment of these controls can create a warm look or a cool look if desired. In this case, let's create a warm look for the image.

Step Nine

First, let's go to the Red tab and adjust the Red Gamma control. The image will become redder through out the image. However, it's not the warm feeling we're looking for at this point.

We then move to the Green tab and adjust the Green Gamma control to create the warm look we want throughout the image.

Step Ten

Now that our correction is completed, we'll click with our mouse on the "OK" button, which closes the Synthetic Aperture window and shows the color correction on the Adobe Premiere desktop display. In order to apply this correction throughout the selected scene, we'll need to render our correction.

Step Eleven

In order to render the information, we select Timeline from the pull-down menu and click with our mouse on Render Work Area. When this is done, the software will prompt us to save our project and proceed with the rendering.

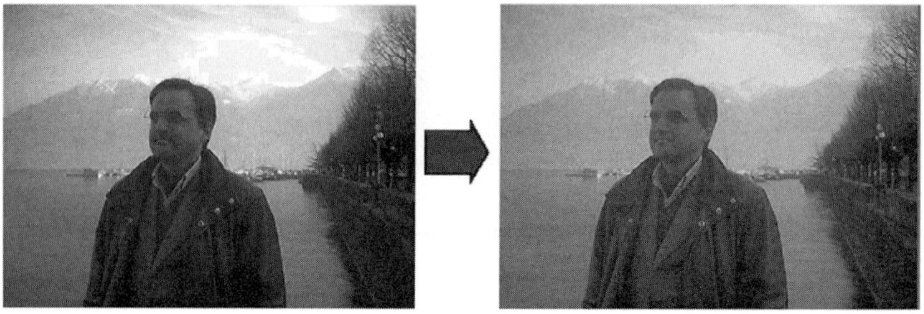

***Image** 7-1 An example of primary color correction.*

Selective Color Correction

In this tutorial, we're going to perform a selective color correction. We'll be doing this using Adobe After Effects version 5.5 along with the Composite Suite Color Correction plug-in from Digital Film Tools. This tutorial can be also seen on the companion CD-ROM.

Step One

Once the clip has been imported into After Effects and has been inserted into the timeline, the next step is to select the method of correction. Using the Effect pull-down menu, select DFT Composite Suite followed by selecting CS Selective Color Correct.

Step Two

Once this is selected with the mouse, the effects control window will automatically open. A mouse click on any of the arrows located to the left of a label within this window will reveal the slider controls. The drop-down menu next to the View label allows us to select output, original, or matte views. We want to select the matte view option by changing the drop-down menu from output to matte.

Step Three

We then click on the arrow to the left of the matte label, which will reveal the various options available. For this exercise, we'll derive the matte based on the hue of a color, so we'll select extract from Hue.

Step Four

Now it's time to select the area within which we want color correct. This is done by clicking on the arrow next to the Position label and adjusting the slider, which will show the outcome of the matte.

Step Five

If the material we're using has some compression in it, we may want to defocus the matte in order to improve the correction. This can be done by clicking on and adjusting the Blur control, which will apply a defocus correction to the matte.

Step Six

Now that the matte is adjusted so that the correct information is revealed, we can change the view control back to Output. It's now time to think about the correction we're going to put in.

Step Seven

To color correct within the region selected with the matte generator, we can now select any of the color correction options within this window, which will change the values within the selected area. If we want to see what the original image looks like without this correction, this can be done by changing the View option from Output to Original.

Step Eight

Now that we're happy with the color correction, we can proceed to rendering it by selecting Add To Render Queue from the Composition drop-down window.

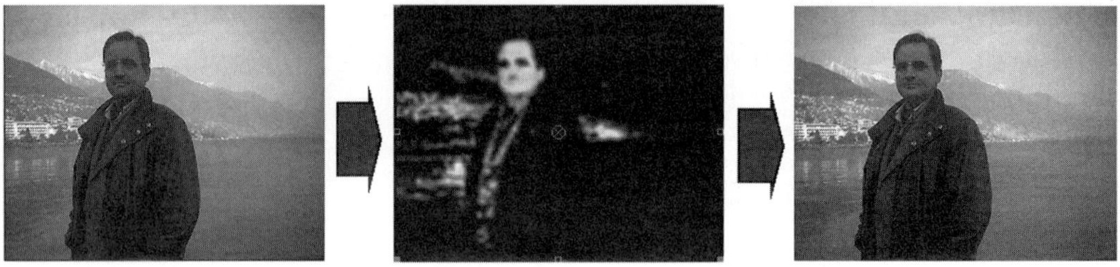

Image 7-2 *A selective color correction. (See also Color Plate 6.)*

Creating a Color Gradation

The color gradation, or "grad" as it's known by among cinematographers, can be easily generated with video color correction. In this tutorial, we'll be doing this with Adobe After Effects version 5.5 along with the Digital Film Tools 55 mm plug-in application. This tutorial can be also seen in the companion CD-ROM.

Step One

We first will import the clip into After Effects, where it will show up in the Project window. We can then place this in the timeline by using the drag-and-drop method. Following this, the content will be displayed on the Composition display.

Step Two

Now, we can click with the mouse on the Effect drop-down menu, and we'll follow this up by selecting DFT 55 mm. After this, select 55 mm Color Grad, which will automatically open your Effects Controls window.

Step Three

The gradation automatically appears; however, we would like to customize it. In order to easily do this, we reset the Size control to 1.0 by clicking with the mouse on the blue numbers to the right of the Size label, typing in 0, and then pressing Enter.

Step Four

Clicking with the mouse and moving the slider to change the Position value will alter the size of the colored area. We can then change the gradation color by clicking with the mouse on the color button next to the Tint label. This will open another window, which allows us to select the desired color by holding down the left mouse button on the color display and running it back and forth until the desired color is arrived at. Once this is finished, we click on OK with the mouse, and the window will close, resulting in the selected color appearing within the gradation area.

Step Five

The opacity of the color gradation can be changed by simply clicking with the mouse on the opacity control and moving the slider back and forth to the desired level. Generally, moving the slider to the right will give better results because moving the slider further to the left will make the gradation color almost invisible.

Step Six

Depending on the angle of the horizon, we might want to put a little tilt in the gradation. This can be done by clicking with the mouse on the arrow next to the Rotation label and then clicking on the small rotation circle that appears. Holding down the mouse button and moving the mouse while in this display will rotate the gradation to the desired position.

Step Seven

Now the parameters of the gradation are just about correct, but the whole image, including the gradation, may need an overall correction. Clicking on the arrow next to the Density label will reveal the Brightness, Gamma, and Contrast controls, which can be used to change the levels throughout the picture. This is done by clicking on the desired control and changing the value using the slider control.

Step Eight

We're almost finished, so now it's time to apply some softness, which can be done by clicking on Size and changing the value with the slider control. Generally, just a small change will give the desired effect.

Step Nine

So now we've finished the gradation, and it's time to put it into the Render Queue by selecting Add To Render Queue from the Composition drop-down menu.

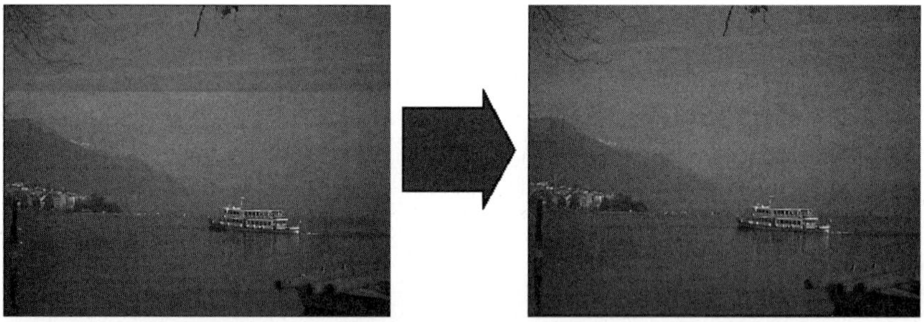

Image 7-3 *Creating a color gradation.*

Creating a Knee Correction

The Knee adjustment is essentially a soft clipping tool. In this exercise, we've already performed primary color correction; however, the whites are a little clipped, and we would like to recover some detail within these areas. For this

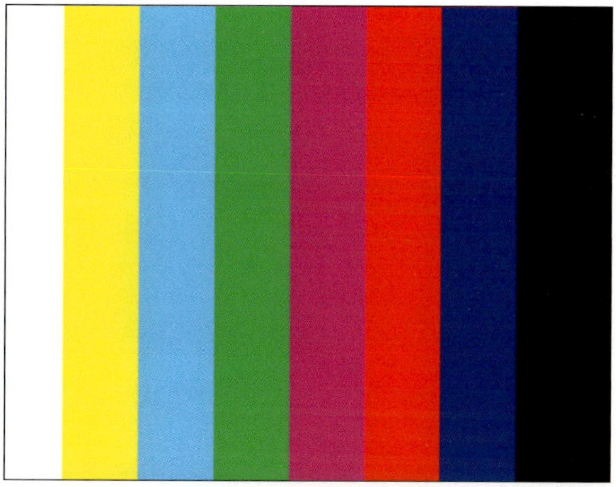

Plate 1 *Full field color bars*

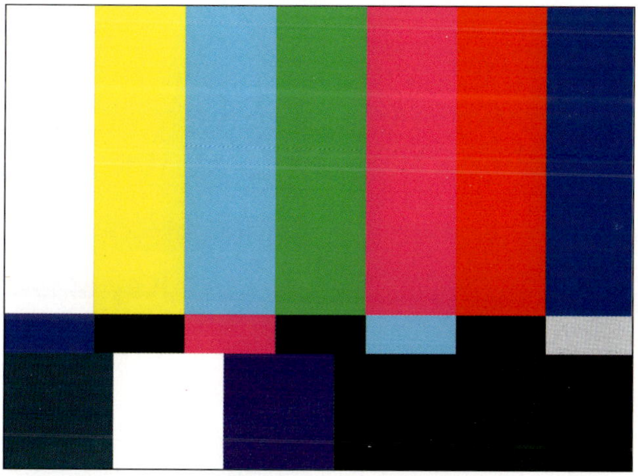

Plate 2 *SMPTE color bars*

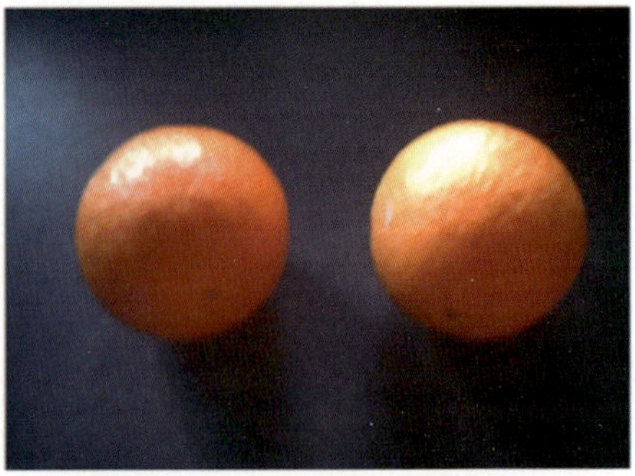

Normal Correction

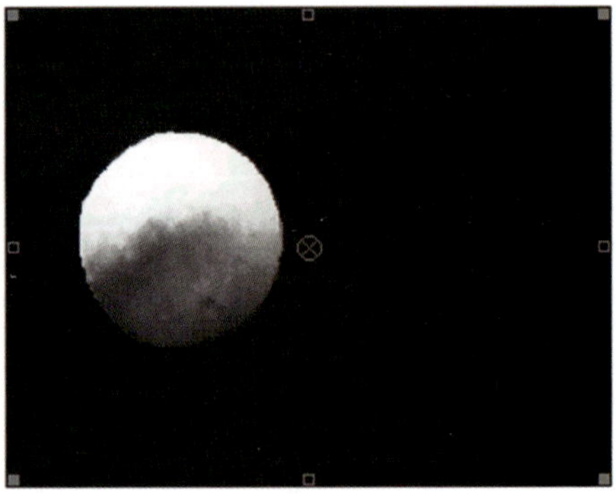

Key Mask

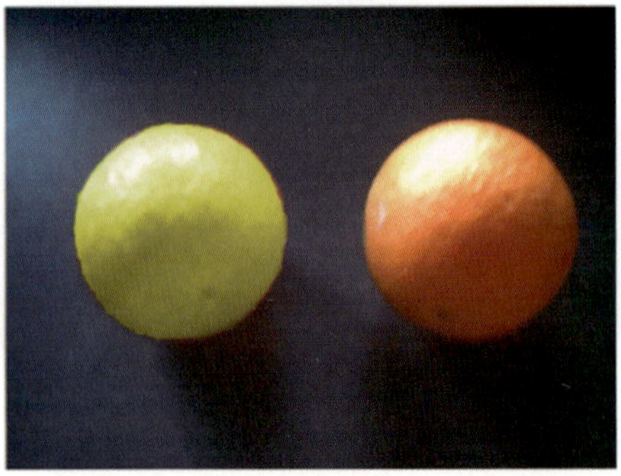

Secondary Color Correction Applied

Plate 3 *Secondary color correction*

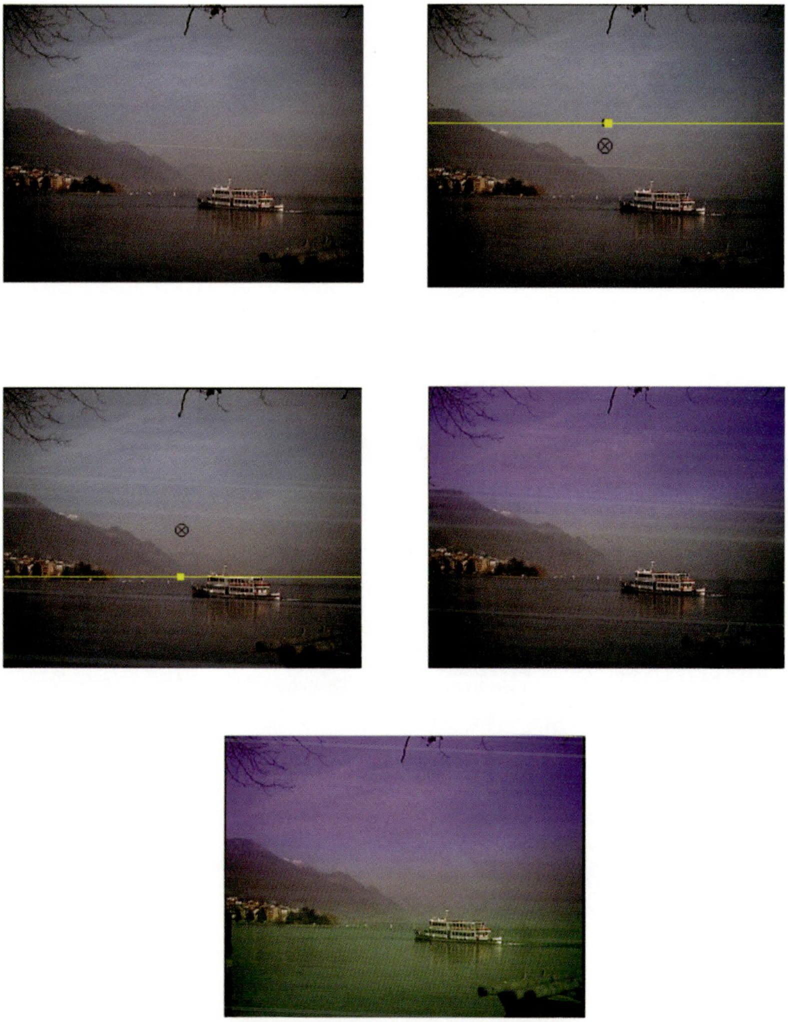

Plate 4 *A multiple gradation*

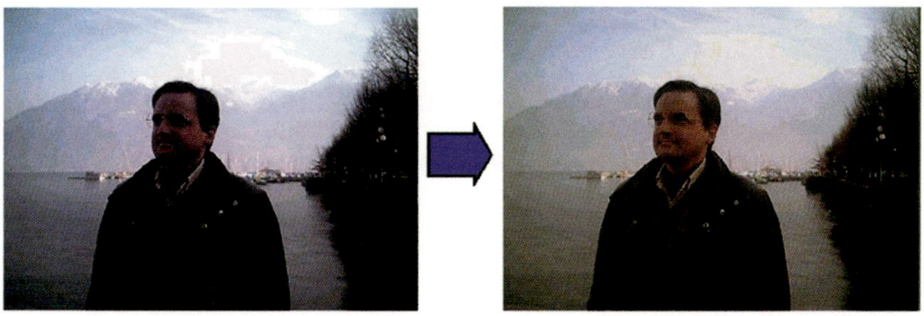

Plate 5 *An example of the Composite Suite color-correction plug-in's primary color corrector*

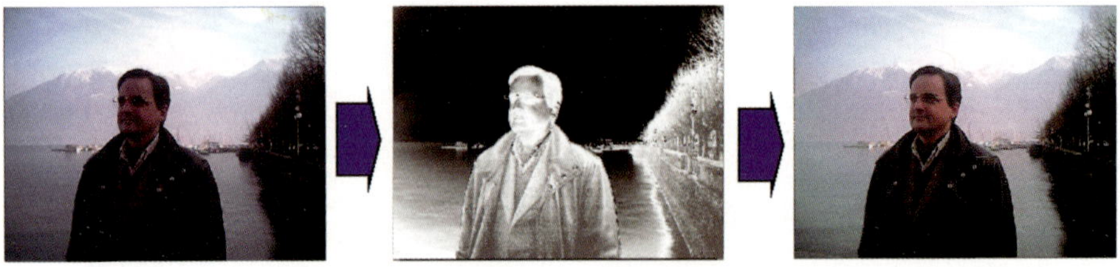

Plate 6 *A selective color correction*

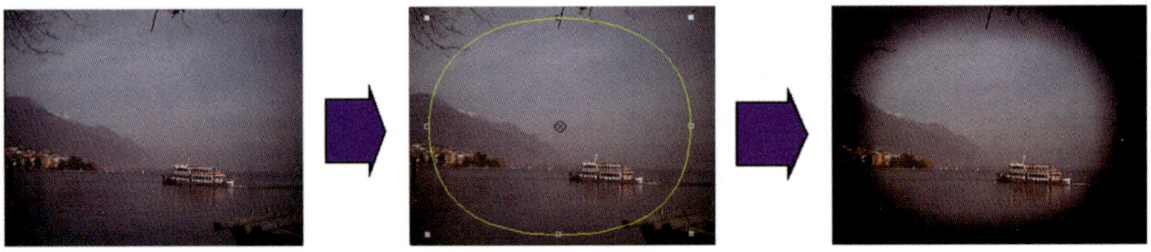

Plate 7 *Creating a vignette*

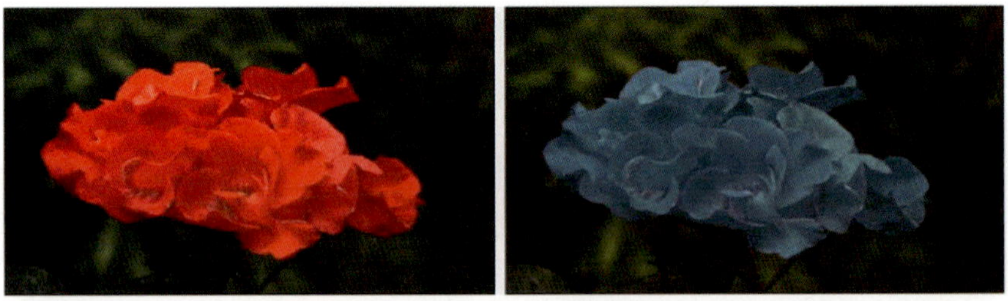

Plate 8 *A secondary color correction achieved with Color Finesse (Courtesy of Synthetic Aperture)*

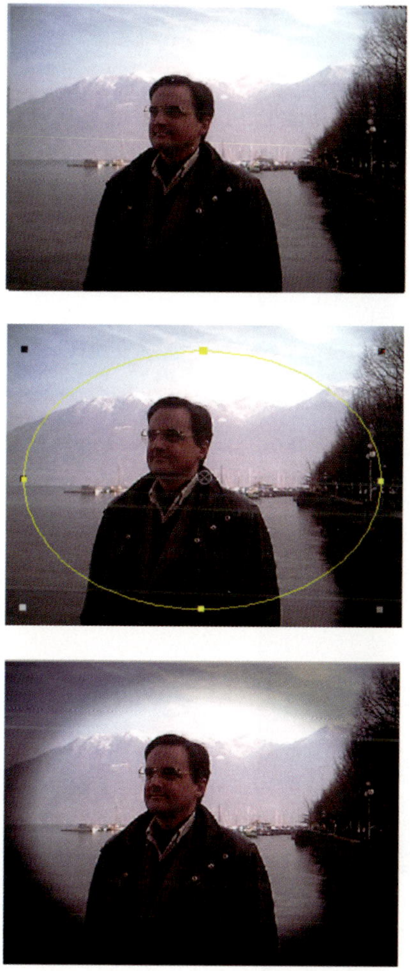

Plate 9 *A colored vignette*

Plate 10 *Creating the night vision effect*

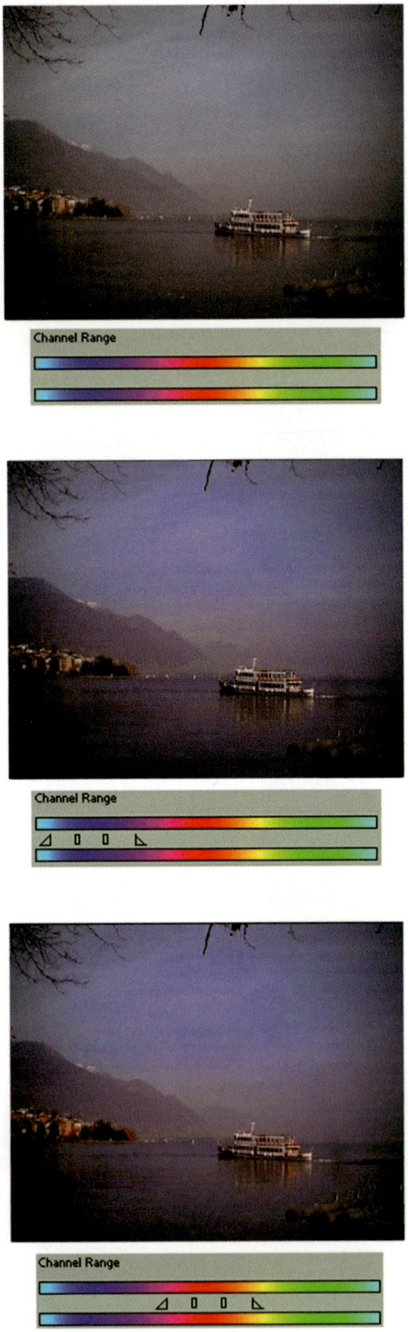

Plate 11 *Multiple secondary color corrections*

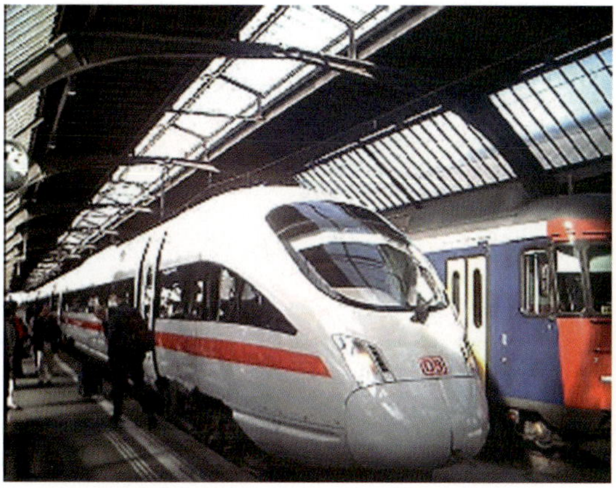

Plate 12 *Uncorrected image*

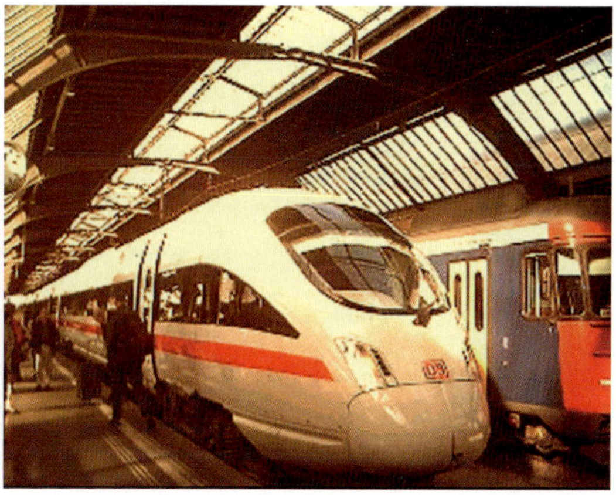

Plate 13 *Warm colored image*

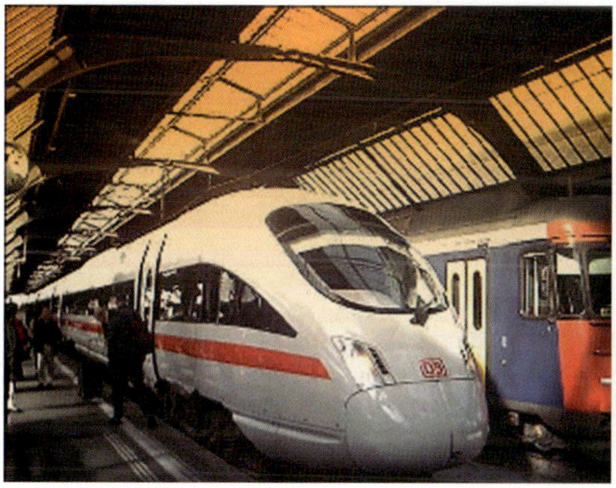

Plate 14 *A yellow gradation*

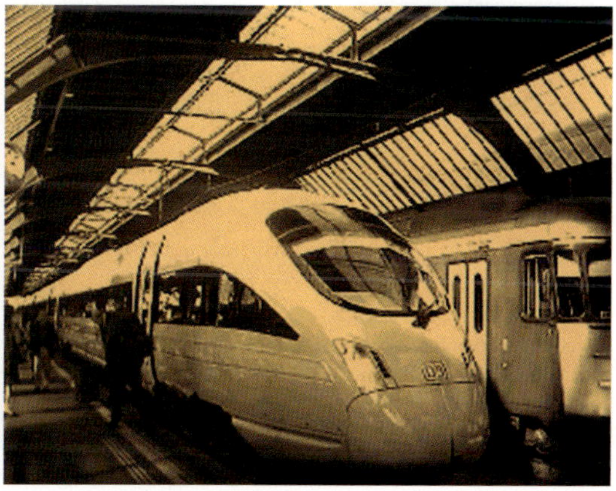

Plate 15 *A sepia tint*

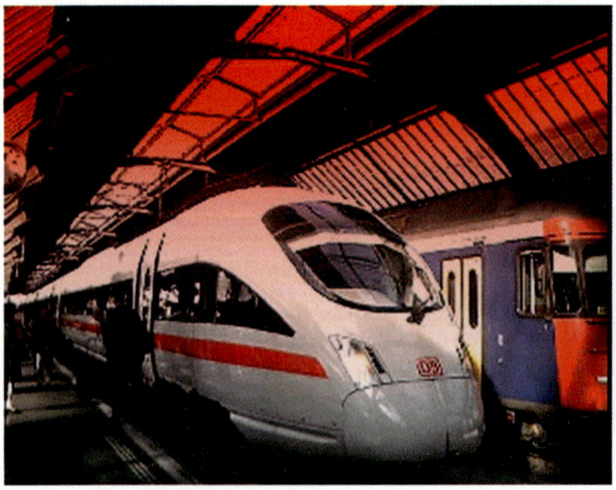

Plate 16 *A red gradation*

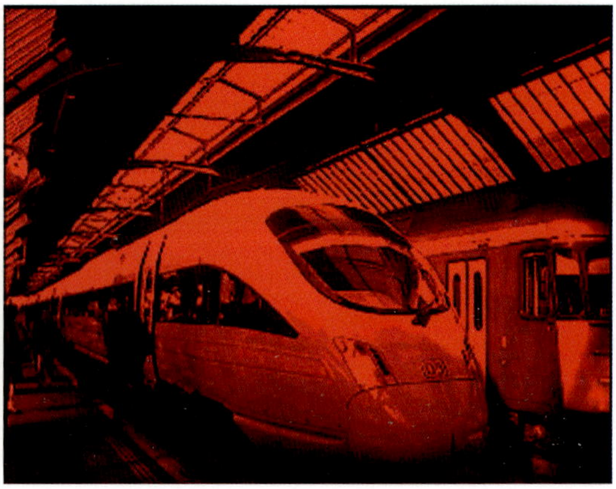

Plate 17 *A red image tint*

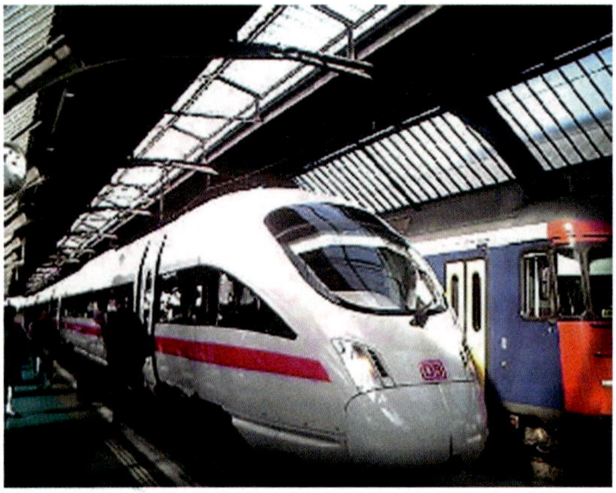

Plate 18 *A selective color correction*

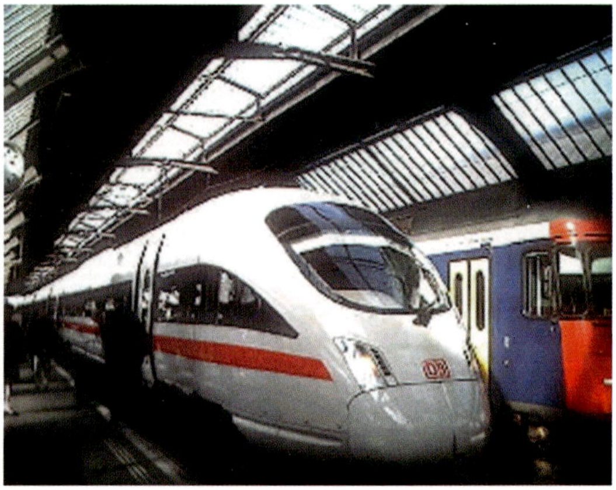

Plate 19 *Selective defocus*

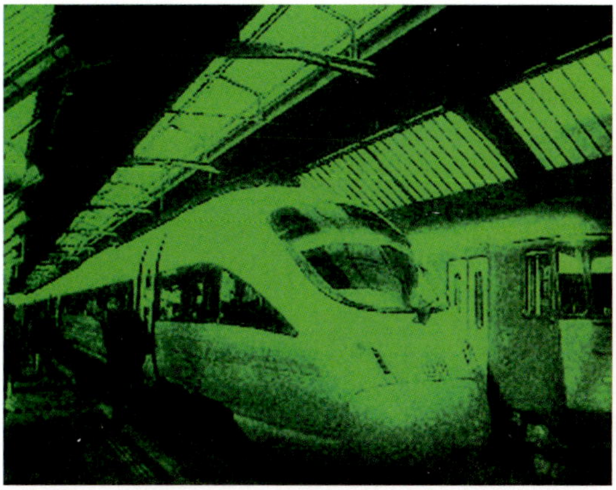

***Plate 20** A night scope effect*

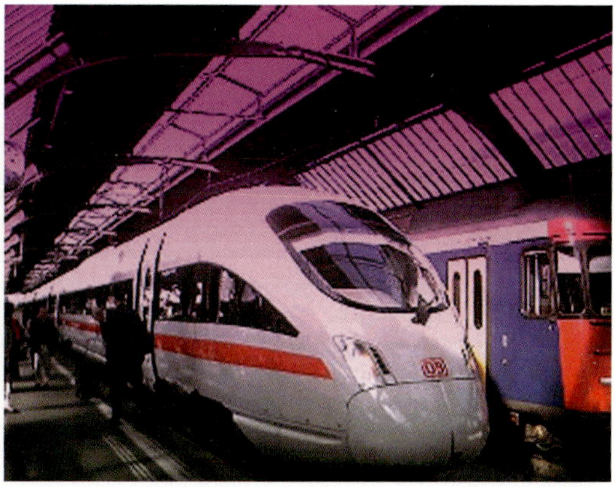

***Plate 21** A pink gradation*

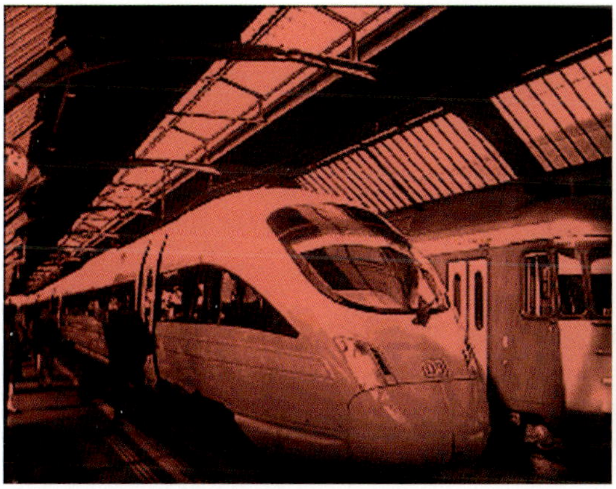

***Plate 22** A pink image tint*

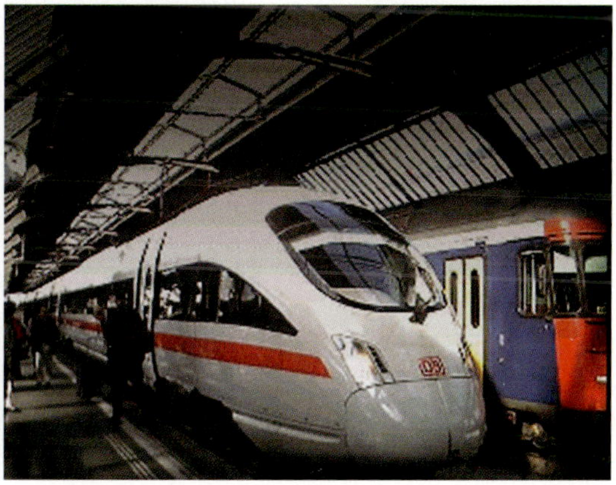

***Plate 23** A neutral density gradation*

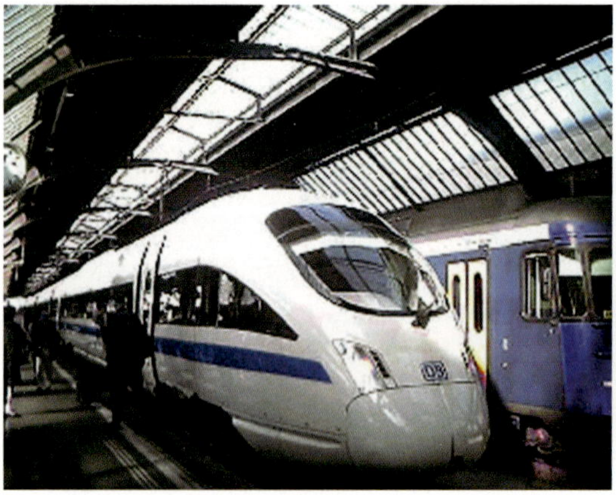

Plate 24 *A secondary color correction*

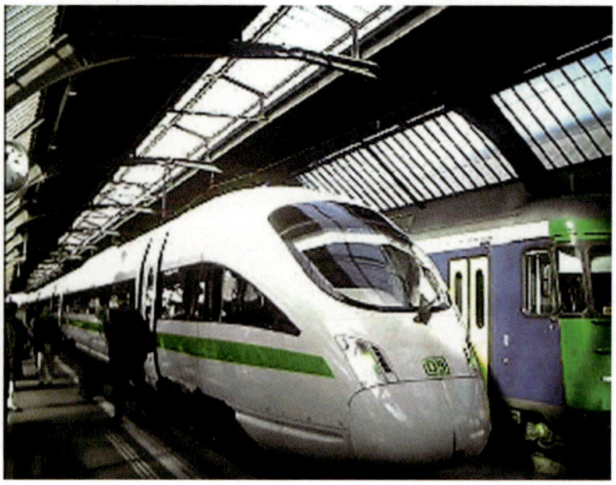

Plate 25 *A secondary color correction*

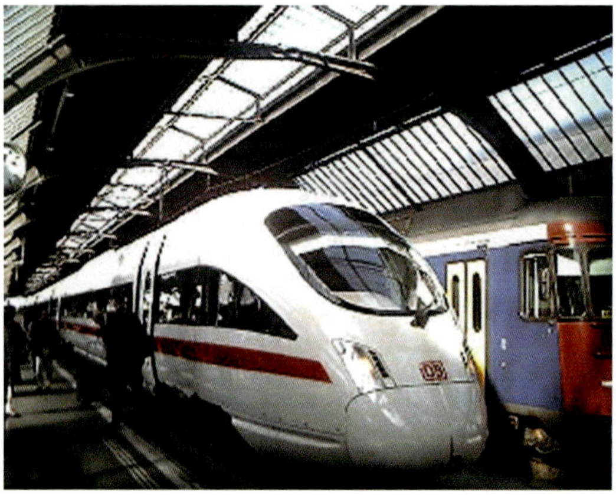

Plate 26 *A secondary color correction*

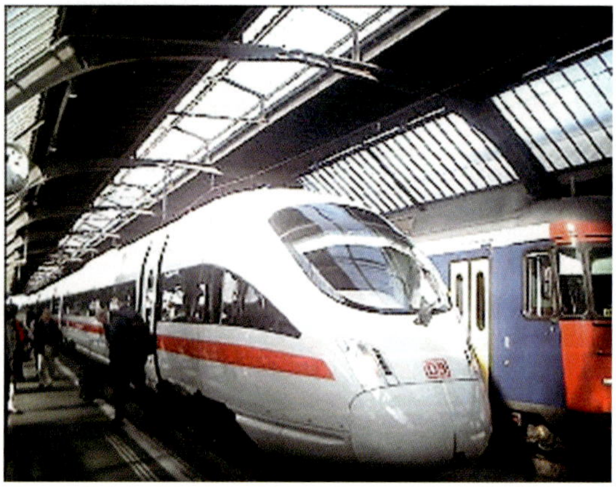

Plate 27 *A fog effect*

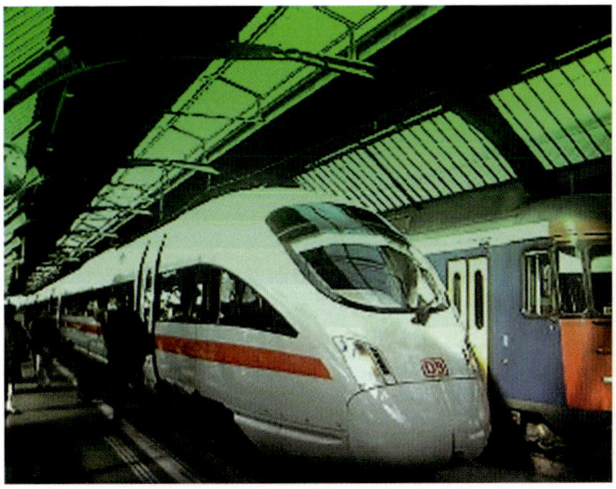

Plate 28 *A green gradation*

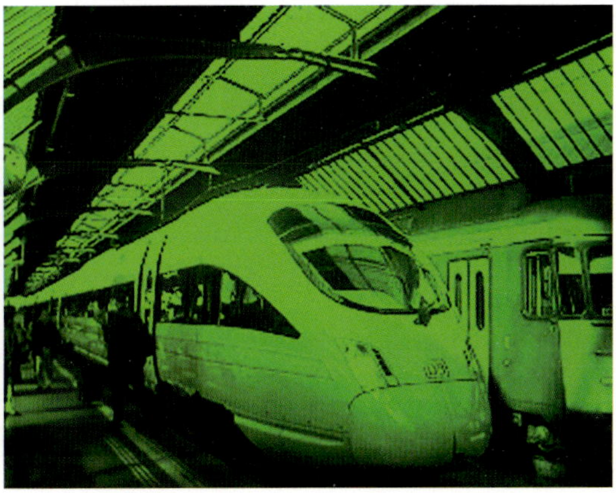

Plate 29 *A green image tint*

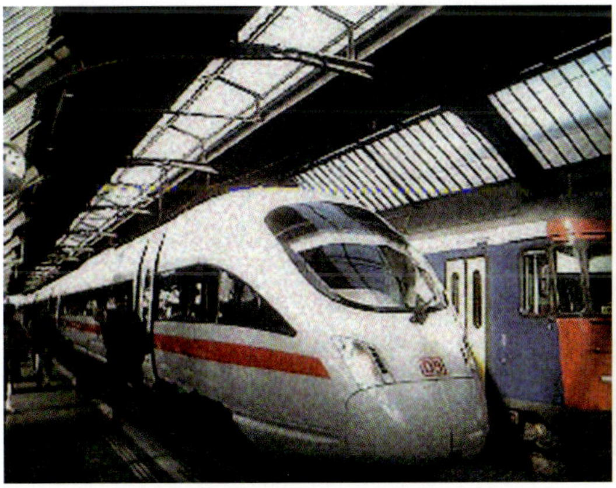

Plate 30 *The faux film effect*

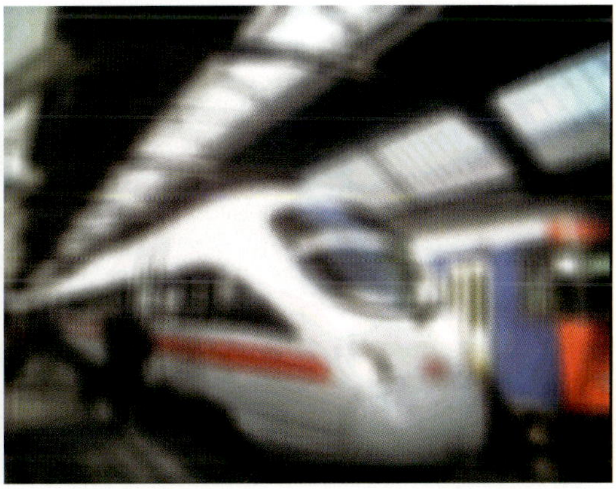

Plate 31 *A defocus effect*

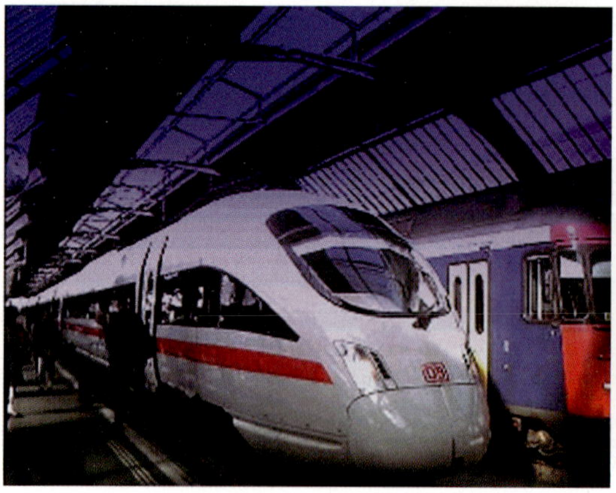

Plate 32 *A blue gradation*

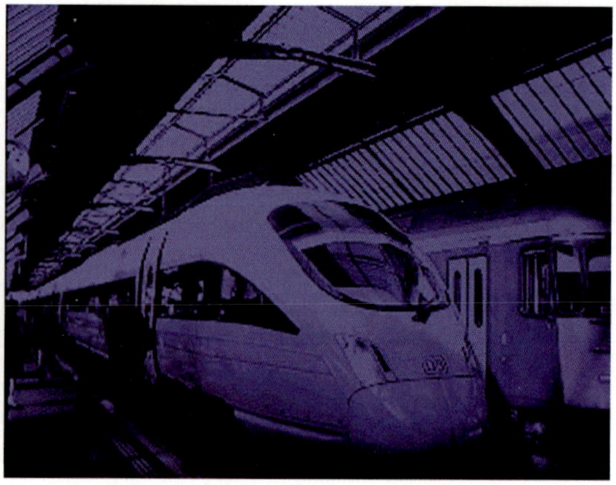

Plate 33 *A blue image tint*

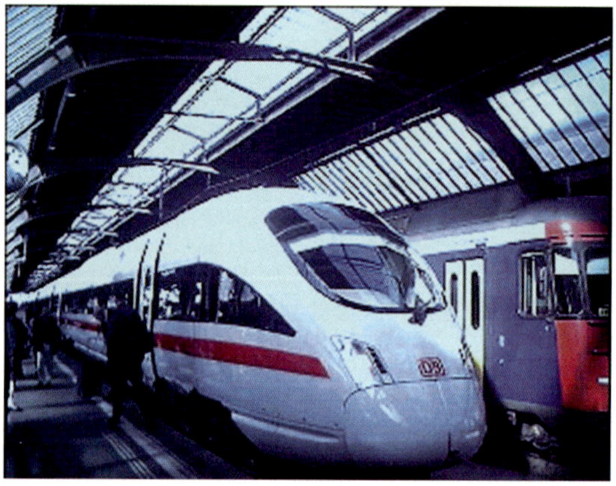

Plate 34 *A cool color correction*

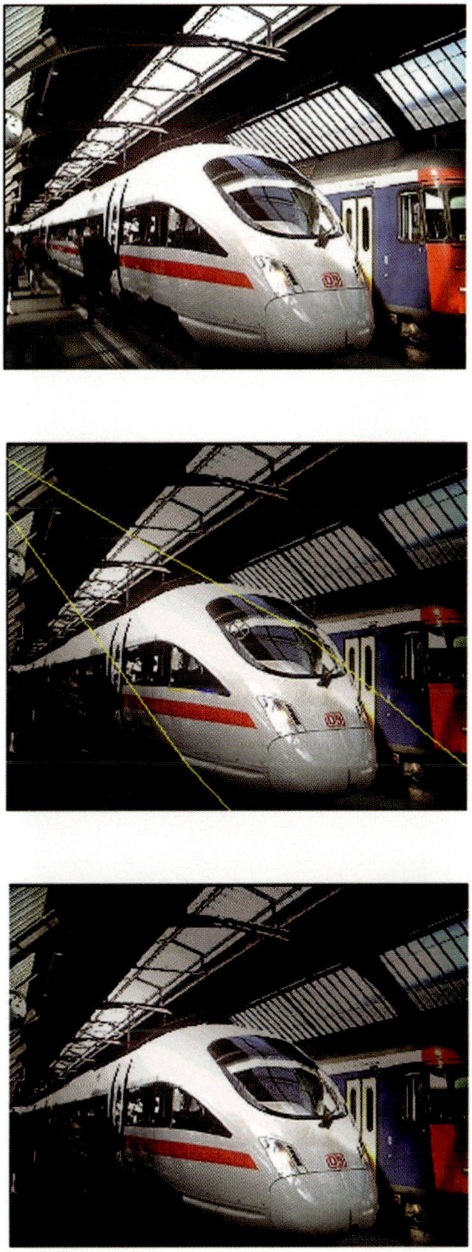

Plate 35 *A shaft of light vignette*

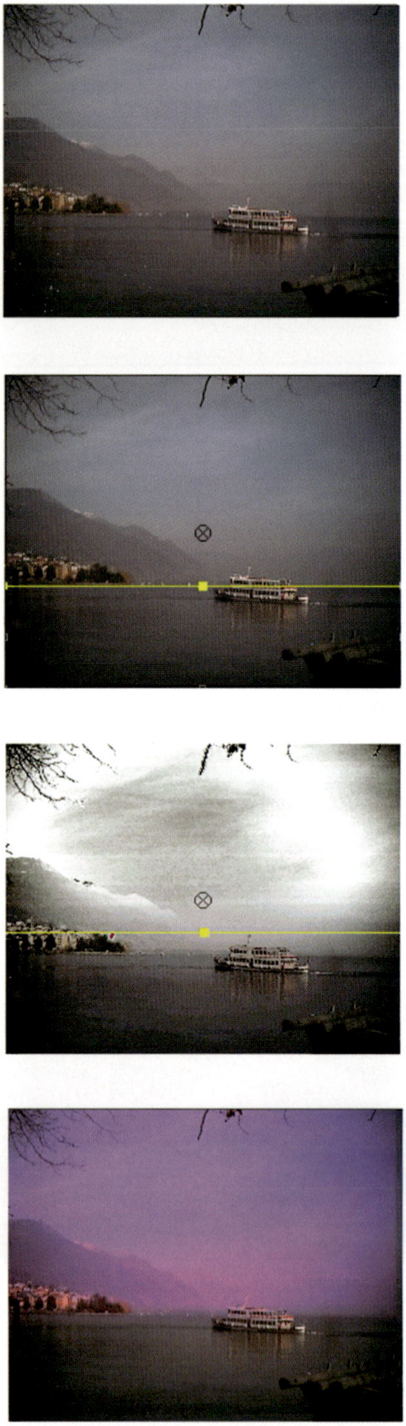

Plate 36 *A selective color correction with a mask*

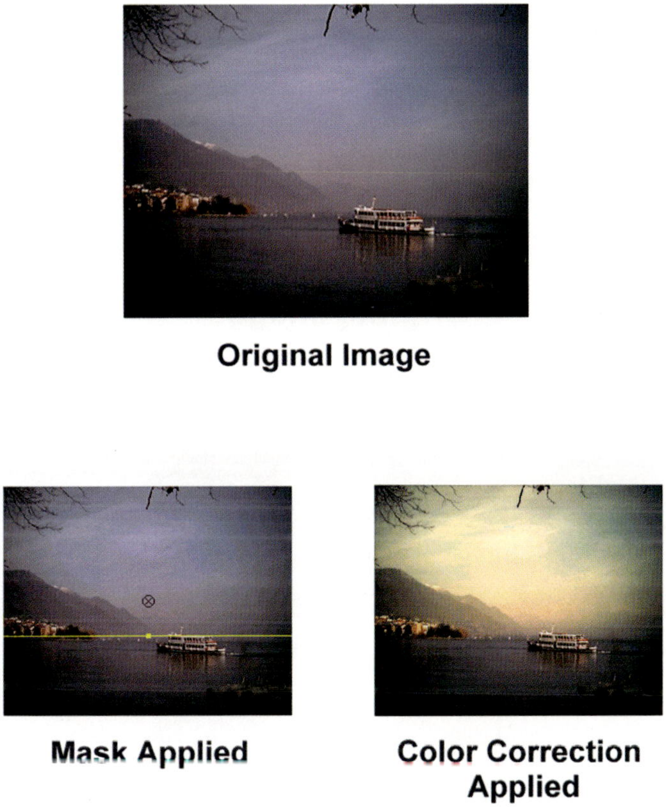

Plate 37 *A masked color correction*

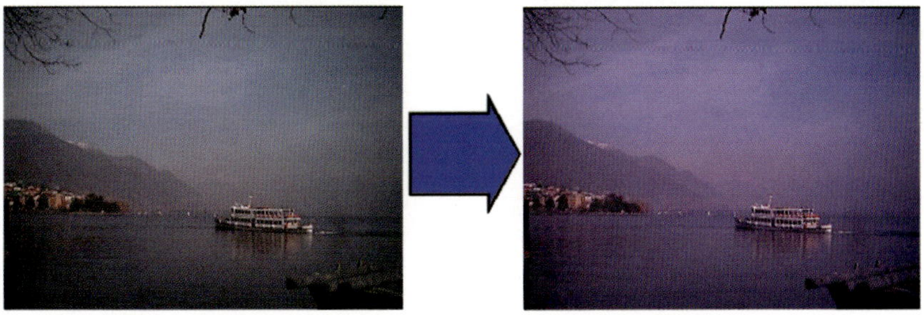

Plate 38 *Creating a dual-tone effect*

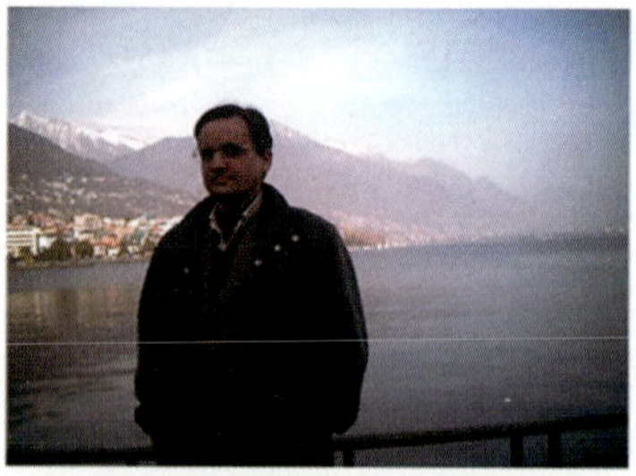

Original Image

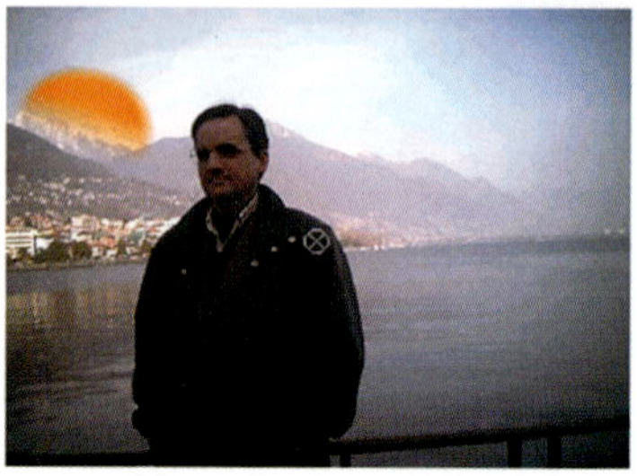

With One Mask

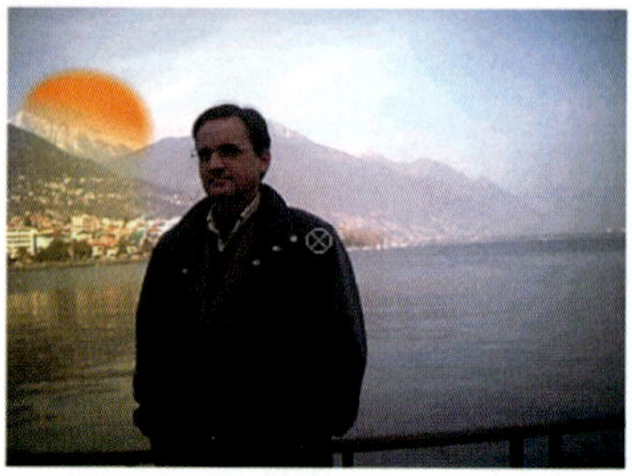

Adding A Second Mask

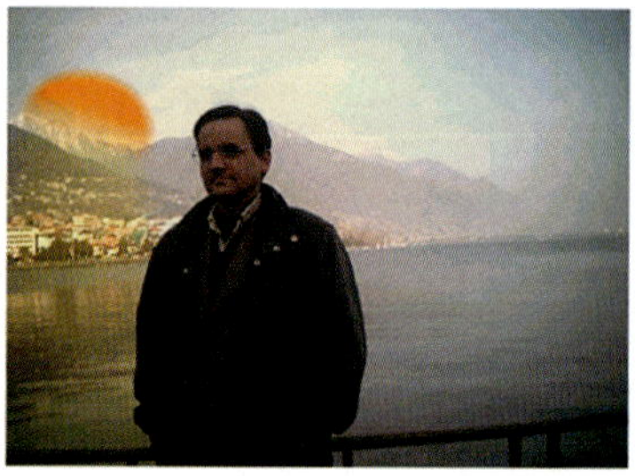

The Finished Result
With An Over All Correction

Plate 39 *Multiple masks within an image*

exercise, we're going to be using Adobe Premiere version 6.0 along with Synthetic Aperture's Video Finesse plug-in application. This tutorial can be also seen in the companion CD-ROM.

Step One

We are assuming that by this stage we've imported the clip into the system and dropped it into the timeline. If the Video Effects window is not visible on the desktop, it can be brought up by selecting Show Video Effects from the Window from the drop-down menu at the top of the screen.

Step Two

We'll now select SA Adjust located in the 3rd Party video effects, hold down the left mouse button, drag it over to the timeline, and release the mouse button. Following this, the Synthetic Aperture window should automatically appear.

Step Three

If we click with the mouse on the Knee Level control and adjust the slider, we can select the amount of soft clipping. The Knee Slope adjustment will adjust the slope for the rounding of the soft clipping. If the knee level is at 100, the slope control will have no effect at all.

Step Four

If we now bring the Knee Level control about halfway down and adjust the slope, we have a good starting point at which to begin. Synthetic Aperture provides a nice little ramp that displays the curve that is created. You pretty much want to end up with as rounded a curve as possible in order to achieve a nice soft clipping effect.

Step Five

Once the level is set to our satisfaction, we adjust the slope to recover the lost information in the highlights. If we feel that not enough information is coming back, we can return to the Knee Level control and touch up the adjustment a little bit.

Step Six

At this point, we're pleased with our knee adjustment, so we click with the mouse on OK, and the Synthetic Aperture window automatically closes.

Step Seven

We're just about done, so we select Render Work Area from the Timeline drop-down menu located at the top of the desktop display.

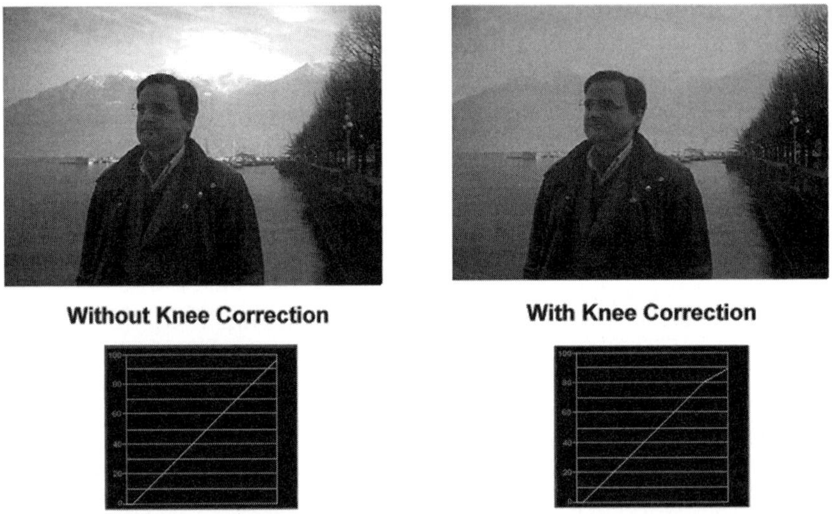

Image 7-4 *A Knee control adjustment.*

A Defocused Halo Effect

This type of effect can be extremely useful in certain types of situations, such as music videos, fashion material, and so on. This exercise is being carried out using Adobe After Effects 5.5 along with Digital Film Tools' Composite Suite Color Correction plug-in application. This tutorial can also be seen on the companion CD-ROM.

Step One

First, we'll import the clip into After Effects by selecting Import from the File drop-down menu. The clip appears in the Project window.

Step Two

Now, we'll drag the clip from the project window and drop it onto the timeline so that we can proceed with our correction.

Step Three

Now, we'll select our correction tool by selecting DFT Composite Suite from the Effect drop-down menu, followed by selecting CS Selective Color Correct.

Step Four

Once Step three has been completed, the Effects Controls window will auto matically open, revealing the adjustment controls. We select Matte on the View drop-down window and see the matte signal.

Step Five

Clicking on the arrow next to the matte label reveals the controls for adjusting the matte signal's properties. Next, we want to extract a matte signal based on the luminance within the picture, so we'll change Extract On to Luminance on the drop-down menu located within the Matte Selection area.

Step Six

Now, we'll adjust the area to be affected by changing the Position control, which is done by adjusting the slider with your mouse. If required, the Range control can be slightly adjusted by using the slider control.

Step Seven

Following this, let's apply a defocus effect to the matte by clicking on the Blur label with our mouse and moving the slider control. When the slider is positioned all the way to the left, the image is sharp, and when it's positioned to the right, the image is defocused. In this situation, we just want to defocus it a little bit, so we position the slider about one-third of the way over.

Step Eight

Next, we adjust the Shrink/Grow control in order to have a nice clean matte signal that is nicely defocused around the edges.

Step Nine

Our matte is all set up, so we change the View drop-down menu within the control window to Output.

Step Ten

Now, let's change the levels and color within the area by using the CS Selective Color Correct tools located in the lower part of its display within the Effects Control window.

Step Eleven

A small adjustment to the Brightness control instantly reveals our halo effect. Now, we'll put a bit of color correction in there by adjusting either the Red, Green, or Blue color controls. The result will be a colored halo around the subject.

Step Twelve

We're now happy with our correction, and it's render time! We can proceed to this by simply selecting the Composition drop-down menu located at the top of the main After Effects desktop, followed by selecting Add To Render Queue.

The Render Queue window will open, and clicking our mouse on Render will start the process.

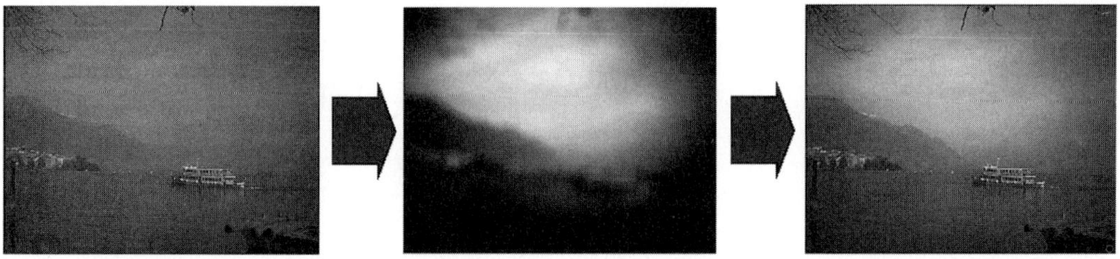

***Image** 7-5 Applying a defocused halo effect.*

Creating a Night Scope Color Effect

When you watch the news, it's almost a daily occurrence that some material in the field has been shot through a night scope. Well now it's easy to create this sort of effect. In this exercise, we'll be doing this by using Adobe After Effects along with the Digital Film Tools 55 mm plug-in application. This tutorial can also be seen on the companion CD-ROM.

Step One

Our first step is to import the clip into After Effects. Once this is done, the clip will show up in our project window.

Step Two

We'll then drag the clip from the Project window and drop it onto the timeline. The image will now be seen in the After Effects display.

Step Three

We now want to select the appropriate tool by clicking on the Effects drop-down menu located at the top of the After Effects desktop and selecting DFT 55 mm. While in this menu, we select 55 mm Night Vision.

Step Four

Once Step Three has been successfully completed, the Effect Control window will automatically open, revealing the controls for customizing the look of the image. Once this is done, the image on the After Effects display will turn green.

Step Five

Following this, we'll adjust the Brightness control by clicking the arrow next to the label and adjusting the slider to add a glow to the image.

Step Six

Next, we want to blur the highlights a bit by clicking on the Blur control and adjusting it with the slider.

Step Seven

We're almost ready; however, we want to add a little grain in order to make the image look like the material we saw on the news. We do this by clicking on the arrow next to the Grain label, which reveals the other grain customizing controls. Next, we'll adjust the amount of grain to the desired amount, which is usually done by moving the slider about a quarter of the way.

Step Eight

We're now happy with our night scope effect, and it's render time! We simply select the Composition drop-down menu located at the top of the main After Effects desktop and then select Add To Render Queue. The Render Queue window opens and clicking on Render will start the process.

Image 7-6 *Creating a night vision effect. (See also Color Plate 10.)*

Using Adjustable Curves

Adjustable curves can be extremely helpful in a variety of situations. In this tutorial, we'll be using the adjustable curve tool contained within Adobe After Effects version 5.5. This tutorial can be also seen on the companion CD-ROM (using a different image).

Step One

First we want to import our clip into After Effects by selecting Import from the File drop-down menu located at the top of the After Effects desktop display. After importing the clip, it should be visible in the Project window.

Step Two

Our next step is to drag and drop the clip into the time line. When this has been successfully completed, the image will be seen in the After Effects display area.

Step Three

Now, we want to bring up the correction tool. This is done by selecting the Effect drop-down menu, followed by selecting the Adjust menu. Once we have located the Adjust menu, we click on the submenu labeled Curves.

Step Four

Once the Curves tool has been selected, the Effect Control window will open automatically, revealing the curve graph. The top right of the graph indicates peak white, while the bottom left indicates the black information. There are many ways to modify this graph. One is to select the pen option and draw a graph using your mouse. The easiest method is to simply select the lines with the points option. When this is selected the operator merely modifies the graph by adjusting the points with the mouse. A click with the left mouse button will add points, while moving the mouse and holding down the left mouse button will modify the graph. Whenever you feel that your correction is not appropriate, it can be reset by clicking the Reset button at the top of the control window.

Step Five

We want to create an overall curve, so we'll select RGB, which is located next to the channel label. We could select Red, Green, or Blue, but these options would only perform the correction in their respective channels.

Step Six

For this particular tutorial, we want to create an "S" style RGB curve in order to simulate a film print look. This can be simply done by adding two points to the graph, one at the lower-left corner of the top right block within the graph. The next point should be made at the upper-right corner of the lower left block in the curve graph.

Step Seven

Next position your mouse over the point created in upper portion of the graph. While holding down your left mouse button, move your mouse in order to adjust the graph by moving it slightly upwards.

Step Eight

Our next step is to position the mouse over the point in the lower part of the graph and hold down the left mouse button. Following this action, move your mouse slightly downward, which will result in the completion of the S curve. The purpose of these two actions is to compress your black information and stretch out your luminance information.

Step Nine

Our curve adjustment is complete, so now we can close the Effect Control window and examine the picture to make sure that we're pleased with the outcome.

Step Ten

We're quite pleased with our curve, and it's render time! We'll render the image simply by selecting the Composition drop-down menu located at the top of the main After Effects desktop, followed by selecting Add To Render Queue.

Following this, the Render Queue window will open, and clicking on Render will start the process.

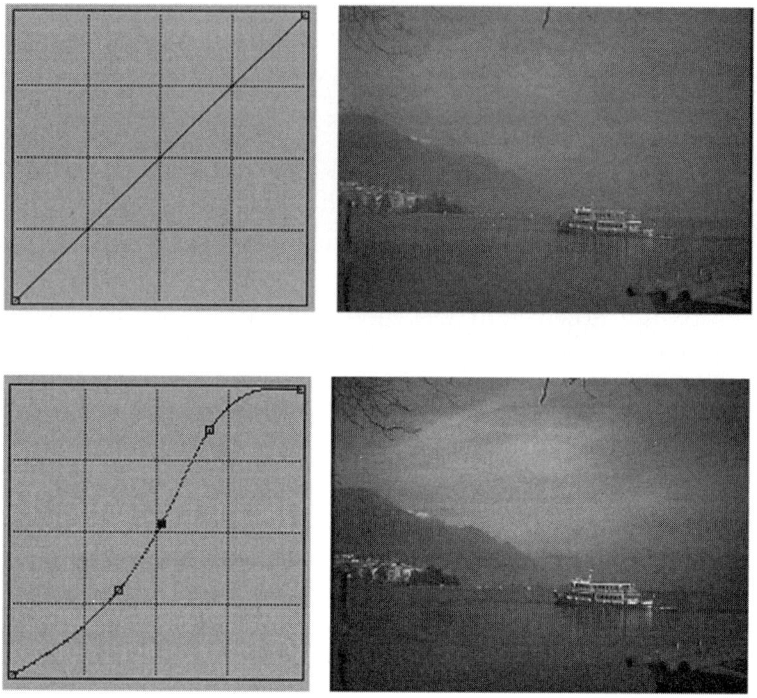

***Image** 7-7 Creating an Adjustable Curve.*

Creating a Secondary Color Correction

In this tutorial, we'll be performing secondary color correction. This will alter the levels of selected colors within the picture. For example, if you select Red, you may change the hue, saturation, and luminance properties of this color. In this particular tutorial, we'll be using the secondary color corrector contained within Adobe After Effects. Additionally, this tutorial can be seen on the companion CD-ROM.

Step One

First, we want to import our clip into After Effects by selecting Import from the File drop-down menu located at the top of the After Effects desktop display. After the clip is imported, it should be visible in the Project window.

Step Two

Our next step is to drag and drop the clip into the timeline. When this has been successfully completed, the image will appear in the After Effects display area.

Step Three

Now, we want to bring up the correction tool. This is done by selecting the Effect drop-down menu, followed by selecting the Adjust menu. Once the Adjust menu has been opened, we click Hue/Saturation.

Step Four

Once the Hue/Saturation control has been selected, the Effect Control window will automatically appear. We then click on the arrow next to the label reading channel range. This will reveal the color spectrum and the color vector. If Reds are selected in the Channel Control area, the color display will show the vector positioned over the Red area.

Step Five

For this exercise, we'll select Reds in the channel control area. Next, we want to change the hue of the reds by clicking on the small circle that appears after we click on the arrow next to the Red Hue label. If we hold down the left mouse button and move the mouse left or right, the hue of the reds will be changed. This time, we'll just move it about 36 degrees to the right.

Step Six

Next, we'll want to adjust the red saturation, this is done by clicking on the arrow located next to the Red Saturation label, which will show the adjustable slider. We'll just move the slider about one-quarter of the way to the right.

Step Seven

Now, let's adjust the luminance control. This is accessed in the same way as the saturation control was accessed, except that we're selecting the luminance slider adjustment. In this case, we'll move the slider about one-third of the way to the right.

Step Nine

Our secondary color correction is now complete, so we can close the Effect Control window and make sure we're pleased with the outcome.

Step Ten

We're very happy with the secondary color correction, and it's time to render it. We'll proceed to this simply by selecting the Composition drop-down menu located at the top of the main After Effects desktop display, followed by selecting Add To Render Queue. Following this, the Render Queue window will open, and clicking on Render will start the process.

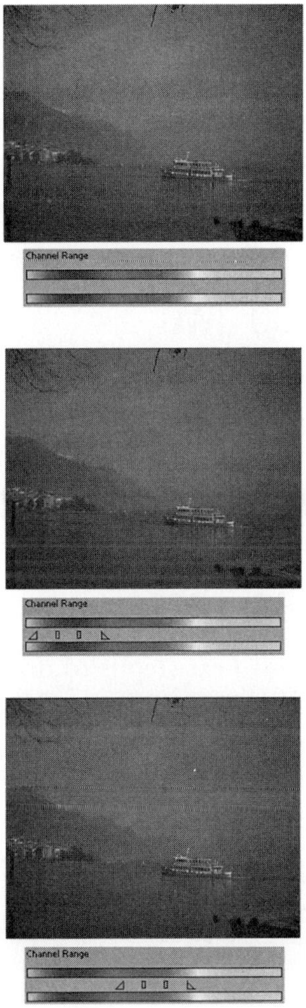

Image 7-8 *A Secondary Color Correction. (See also Color Plate 11.)*

Adjusting Image Levels Using the Histogram

Another method for adjusting the levels of an image is by means of the histogram. The histogram window can be found in most of the applications described in this book. However, on this occasion, we'll be performing the tutorial using Adobe Premiere version 6.0. This tutorial can also be seen as a QuickTime movie on the companion CD-ROM.

Step One

We're assuming that we've imported the clip into the system and dropped it into the timeline. If the Video Effects window is not visible on the desktop, it can be brought up by selecting Window from the drop-down menu at the top of the screen and selecting Show Video Effects.

Step Two

Once Step One has been successfully completed, it's time to choose our correction tool. Upon looking at the Video Effects window that we've just brought up, we see a directory titled Adjust. Clicking on this label will reveal the tools contained within it. For this tutorial, we'll utilize the Levels tool.

Step Three

Our next step is to drag and drop the Levels tool into the timeline. When this is done, our Effect Controls window will open up. At first glance, there doesn't seem to be anything to adjust. We click on Setup, located to the right of the Levels label, and this reveals the adjustable histogram.

Step Four

Located just below the histogram display is series of three arrows. The arrow on the left adjusts the darker areas of the picture, the middle adjusts the mid-tones, and the one on the right adjusts the higher luminance levels. First, let's adjust the darker areas by clicking on the control at the lower left of the histogram display and, while holding down the left mouse button, moving the arrow to the desired position. It's important to observe that the middle arrow is also moving. This effect will also occur when the arrow at the right is adjusted.

Step Five

Once the darker levels in the histogram are adjusted, we now move on to adjusting the lighter parts of the picture by clicking on the arrow located at the lower-right side of the histogram display and moving the arrow with the mouse while holding down the left mouse button. When adjusting both the light and dark areas, it's important to ensure that the resulting picture does not have crushed blacks or clipped whites.

Step Six

At this point, we've adjusted both the light and dark areas. It's now time to adjust the mid-tones within the histogram. We'll now click on the center arrow located just below the histogram display, in the same fashion as we did with the other controls. Next, we'll adjust the control so that we end up with a nice rich-looking picture as seen on the image preview located within this window.

Step Seven

We're now finished with adjusting the histogram display, and the image looks just about right. Located at the bottom of the window is a gray scale display with two arrows, one on either side.

We can adjust these in order to modify the output levels slightly. Because this particular control is downstream, we won't be able to recover any clipped or crushed information created using the other histogram controls.

Step Eight

Now that we're finished with these adjustments, we click on OK to close this window, and we can now give the image a last check to make sure we're happy with the results.

Step Nine

We're quite pleased with the results, so we select Render Work Area, which is located in the Timeline drop-down menu at the top of the Adobe Premiere's desktop display.

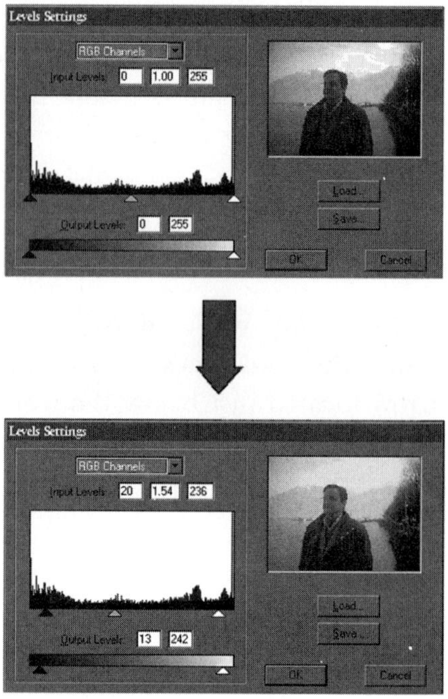

***Image** 7-9 Adjusting Levels Using the Histogram.*

Changing Color Using the Eye Dropper

This feature is available in most software applications and allows the user to choose the color value by using the Eye Dropper. In this application, the Eye Dropper will sample the color value and create a matte based on this color. For this tutorial, we'll be using Adobe After Effects to generate this correction.

Step One

First, we want to import our clip into After Effects by selecting Import from the File drop-down menu located at the top of the After Effects desktop display. After importing the clip, it should be visible in the Project window.

Step Two

Our next step is to drag and drop the clip into the time line. When this has been successfully completed, the image will be seen in the After Effects display area.

Step Three

Now, we want to bring up the correction tool. This is done by choosing the Effect drop-down menu, followed by selecting the Image Control menu. Once the Image Control menu has been located click Change Color.

Step Four

Once Step Three has been successfully completed, the Effect Controls window will automatically open. If we move down the control labeled Color To Change, we'll see a color patch along with an eye dropper icon next to it. So now let's click on the eye dropper.

Step Five

Our next move is to move the mouse around in the picture. Notice how the color patch next to the eye dropper icon in the Effect Controls window is constantly changing during this action. So now we'll click on the area that we wish to change.

Step Six

Next, we'll click on the drop-down menu located next to the View label in the Effect Controls window. The default position for this is to show Corrected Layer. However, let's change it over to Color Correction Mask. Once this has been done, it will show a black and white matte signal, which has been created based on the color chosen with the eye dropper.

Step Seven

We click on and adjust the Matching Softness control using the mouse; this will soften our matte signal.

Step Eight

Next, we can change the tolerance of the matte signal by selecting and changing the Matching Tolerance control.

Step Nine

Our matte signal has been correctly adjusted. We can now change the View menu back to Corrected Layer. Once this has been completed, the picture should return to the After Effects display.

Step Ten

We can now change the color by clicking on and adjusting the sliders associated with the Hue, Lightness, or Saturation Transform controls.

Step Eleven

We're now happy with our correction, and it's render time! We do this simply by selecting the Composition drop-down menu located at the top of the main After Effects desktop, followed by selecting Add To Render Queue. Following this, the Render Queue window will open, and clicking on Render will start the process.

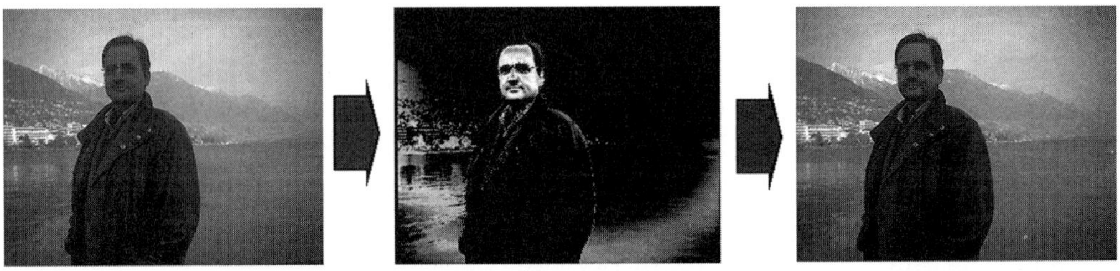

Image 7-10 *Selecting a Color Using the Eye Dropper Tool.*

Creating a Dual-Tone Effect

This type of color effect, as simple as it sounds, was actually quite difficult to create on many stand-alone color-correction computers. Creating this type of effect with the color correction tools in nonlinear applications is extremely simple. The dual-tone effect essentially consists of one color for the dark areas of the picture and one color for the light areas of the picture. This tutorial will be using Adobe After Effects. However, as mentioned, this can be easily achieved on any of the other applications mentioned in this book.

Step One

First, we want to import our clip into After Effects by selecting Import from the File drop-down menu located at the top of the After Effects desktop display. After importing the clip, it should be visible in the Project window.

Step Two

Our next step is to drag and drop the clip into the timeline. When this has been completed, the image will be seen in the After Effects display area.

Step Three

Now, we want to bring up the correction tool. This is done by choosing the Effect drop-down menu, followed by selecting the Image Control menu. Once the Image Control menu has been located, we click Tint.

Step Four

Once we've successfully complete Step Three, the Effect Controls window automatically opens. We'll see two controls, one labeled Map Black To and another labeled Map White To. We can select the colors being mapped into either the black areas or white areas by using the eye dropper function located next to the appropriate control or by clicking on the color patch and choosing the color with the mouse. Once we've chosen the color, we click OK to close the palette window automatically.

Step Five

For this exercise, let's choose Red for the whites and Blue for the blacks. When this has been completed, the colors chosen should show up in the small color patches within the control window.

Step Six

Now that we've successfully chosen our colors, it's time to create the blend. This is done by adjusting the Amount To Tint slider control with the mouse. This will blend the dual tone with the existing colors of the image.

Step Seven

We're now quite pleased with the correction we've made, and it's time to render it! We select the Composition drop-down menu located at the top of the main After Effects desktop, followed by selecting Add To Render Queue. Following this, the Render Queue window will open, and clicking Render will start the process.

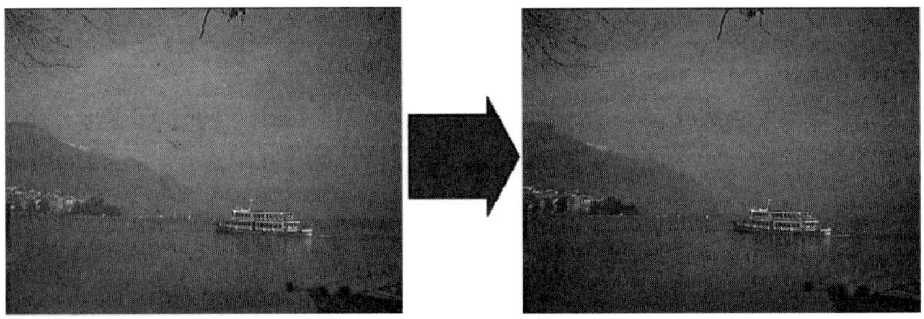

Image 7-11 *Creating a Dual-Tone Effect. (See also Color Plate 38.)*

Instantly Creating a Warm or Cool Look

So here's a real easy one, we're going to make an instant warm or cool look. We'll do this by using Adobe After Effects version 5.5 combined with the 55 mm plug-in from Digital Film Tools.

Step One

After importing the clip into After Effects and inserting it into the timeline, the next step is to select the method of correction. Using the Effect drop-down menu, select DFT 55 mm, followed by selecting 55 mm Warm/Cool.

Step Two

Once this tool has been selected with the mouse, the Effect Controls window will automatically open. We click on the Temperature label; this reveals a slider control.

Moving the position of the Temperature slider control to the left will make the image warmer, and moving it to the right will make the image cooler.

Step Three

For this tutorial, let's make a warmer image. This can be easily done simply by clicking with the mouse on the slider and moving it about one-tenth of the way to the left. We instantly have our warm look.

Step Four

We can compare this with the original color by clicking on the control to the right of the View label within the Effect Controls window. If we change this control to Original, the original color will be shown; changing it back to Output will reveal our correction.

Step Five

After examining the image, we're quite pleased, and it's time to render it. We simply select the Composition drop-down menu located at the top of the main After Effects desktop display, followed by selecting Add To Render Queue. Following this, the Render Queue window will open, and clicking Render will start the process.

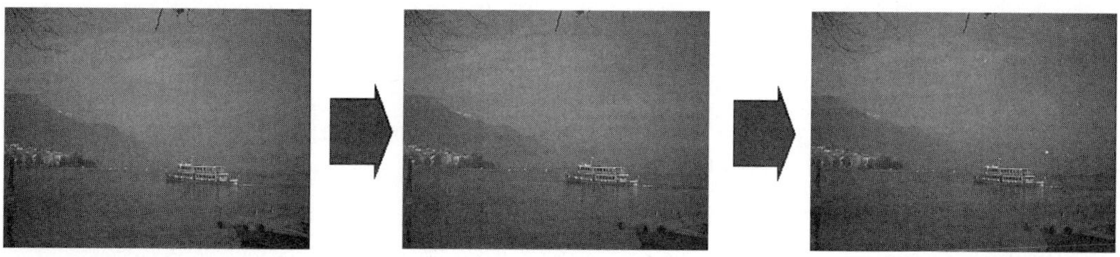

Image 7-12 *Creating an Instant Warm or Cool Look.*

Creating a Black and White Image with One Color Remaining

This is the old trick of keeping one color, while making the rest of an image black and white. We'll be doing this tutorial by using Adobe After Effects' Leave Color tool. Additionally, this tutorial can be seen in the companion CD-ROM (using a different image).

Step One

First, we'll import our clip into After Effects by selecting Import from the File drop-down menu located at the top of the After Effects desktop display. After importing the clip, it should be visible in the Project window.

Step Two

Our next step will be to drag and drop the clip into the timeline. When this has been successfully completed, the image will be visible in the After Effects display area.

Step Three

Next we want to bring up the correction tool. This is done by selecting the Effect drop-down menu, followed by selecting the Stylize menu. After selecting the Stylize menu, click Leave Color.

Step Four

Once this step has been completed, the Effect Controls window will open automatically. We want to now select the color that will remain. This is easily done by clicking on the eye dropper located to the right of the Color To Leave label.

Step Five

We now want to move our mouse to the colored area that should remain and left-click on it. If this has been done successfully, the color patch next to the eye dropper icon will reflect the color chosen.

Step Six

Now, let's change the image into black and white. This is easily done by clicking the Amount To Decolor label and moving the slider to the right. Doing this will remove all the color except the color chosen with the eye dropper.

Step Seven

If there are some areas that are showing color that shouldn't be, these areas can be trimmed by clicking on and adjusting the Tolerance control slider.

Step Eight

If we see a sharp transition point between the color and black and white areas, we can trim this by clicking on and adjusting the Edge Softness control.

Step Nine

We now have a last look at the picture to make sure we're happy with the outcome. If everything is satisfactory, we'll close the Effect Controls window.

Step Ten

Because we're really happy with the correction, it's time to render it. We simply select the Composition drop-down menu located at the top of the main After Effects desktop display. Following this, we'll select Add To Render Queue. Once this has been completed, the Render Queue window will open, and a click with the mouse on the Render button will start the process.

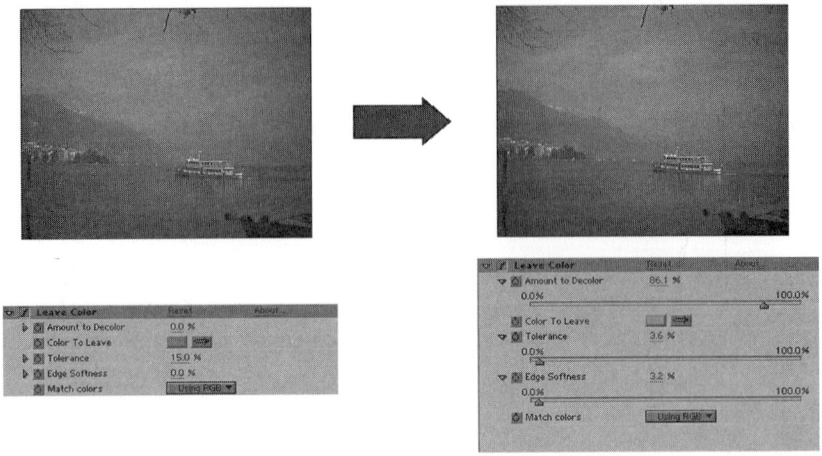

Image 7-13 *Decoloring the Image While Leaving Only One Color.*

Color Correcting Only the Image Highlights

In this exercise, we just want to color correct the highlights of the image, leaving the rest untouched. We'll do this using Adobe After Effects version 5.5 combined with the Composite Suite Color Correction plug-in called CS Selective Color Correct from Digital Film Tools.

Step One

After importing the clip into After Effects and inserting it into the timeline, the next step is to select the method of correction. Using the Effect drop-down menu, we'll select DFT Composite Suite followed by selecting CS Selective Color Correct.

Step Two

After Step One has been completed, we'll see the Effect Controls window, which will automatically appear on the After Effects desktop. The CS Selective Color Correct control menu contains a series of color adjustment controls, along with adjustment controls for the matte isolation signal. We want to start out by isolating the highlights. First, we need to select the drop-down menu located next to the View label within the Effect Controls window. The default position for this control is Output. There is also Original, which is used to compare before and after versions of the color correction. For this example, let's select the Matte option.

Step Three

Once Step Three has been successfully completed, the image on the After Effects display will be replaced with the matte signal. We'll now change the drop-down menu next to Extract On to the Luma position. This will derive a matte signal based on the luminance values within the image.

Step Four

We'll now click on the control labeled Position and adjust the slider control with the mouse. If we move the control to the right, this will isolate the lighter parts of the picture, and moving it to the left will isolate the darker parts. So we'll move the mouse to the right in order to isolate just the highlights.

Step Five

Now that the adjustment to the matte signal has been completed, we can return to the normal image view by changing the View control located in the Effect Controls window from Matte to Output. This action will cause the matte signal being viewed to be replaced with the image.

Step Six

Our next action is to adjust the color of the highlights. This can be easily done by adjusting the controls associated with either the Red, Green, or Blue labels. For this tutorial, let's create a warm look for the highlights. This can be done by clicking on and moving the Red slider about one-third of the way to the right.

Step Seven

At this point, the highlights are quite Red, but we want a warm look. Our next move is to click on and move the Green slider control about a quarter of the way to the right. This addition of green to the Red will create our warm look.

Step Eight

If we wanted to change the overall values within this area, we could also click on and adjust the sliders associated with the Hue, Saturation, and Brightness controls, or even the Contrast or Gamma adjustment controls.

Step Nine

Now, let's have a look at the picture to make sure the result is what we're looking for. If so, we'll close the Effect Controls window and proceed to the next step.

Step Ten

As the correction is fine, it's time to render it. We can start this simply by selecting the Composition drop-down menu located at the top of the main After Effects desktop display. Following this, we'll select Add To Render Queue. After this has been completed, the Render Queue window will open, and we click the Render button to start the process.

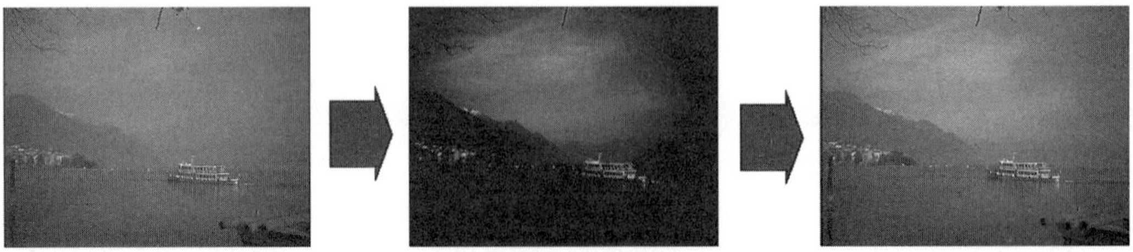

Image 7-14 *An Example of Highlight Color Correction.*

Creating a Sepia Tone Color Correction

This tutorial will be done by using Adobe Premiere 6.0. What we want to accomplish here is to create a black and white image with a sepia tone wash. Upon arriving at Step One, we've already imported the clip we're going to work on, and now we decide that we would like to create our sepia tone correction.

Step One

Our first step will be to bring up the Video Effects window in Premiere. If it's not already on the desktop, it can be opened by selecting Show Video Effects on the Window drop-down menu at the top of the screen.

Step Two

Now, we want to select the tool we're going to use. In this case, we're first going to use the tool labeled Black & White, which is located in the Image Control submenu. We'll select this tool and apply it to the clip by simply positioning the mouse over the tool label. With the mouse still positioned over the label, we'll hold down the left mouse button and drag it over to the timeline.

Step Three

Once the tool has been applied to the scene, the image seen on Premiere will change into black and white.

Step Four

Once our image is in black & white, we want to now introduce our sepia tone correction. We'll select the Color Balance tool, which is located within the Adjust submenu, and apply it to the image.

Step Five

Once this has been completed, the Color Balance controls will appear within the Effect Controls window. We click on the Setup label located to the right of the Color Balance label; this reveals a preview of our image with the color balance controls.

Step Six

So now that we're in this window, if we click on and move the Red slider control slightly to the right, followed by moving the Blue slider control slightly to the left, we should see our sepia effect.

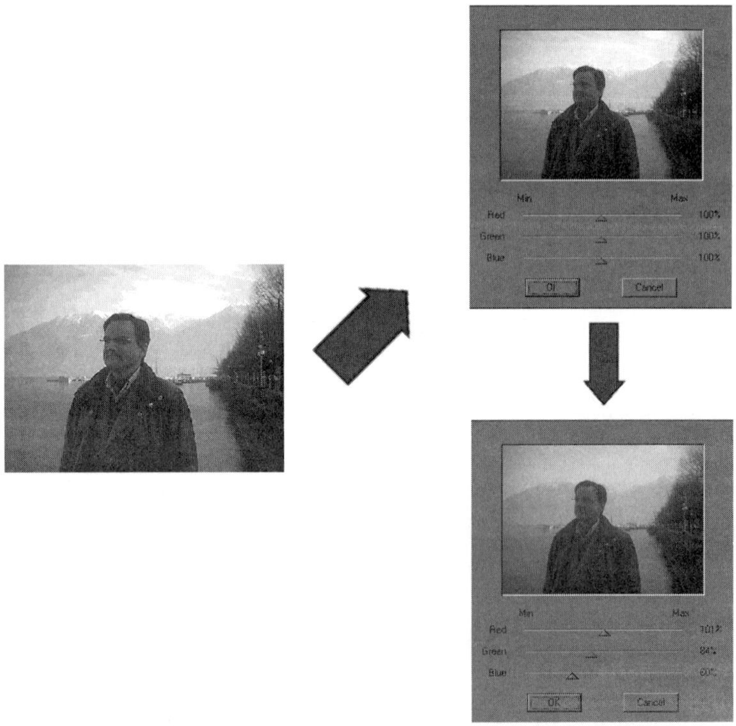

Image 7-15 *Creating a Sepia Tone Effect.*

Step Seven

Once we're happy with this effect, we can close this window by clicking OK. This will close the window, and the correction will appear in Adobe Premiere's image display.

Step Eight

We've created our sepia effect, and now it's time to render it. In order to render the information, we select Timeline from the drop-down menu, and click with the mouse on Render Work Area. When this is done, it will prompt us to save our project and then proceed with the rendering.

Creating a Color Vignette Around an Image

This is a very common effect seen in photography and is also frequently applied when performing color correction in dedicated color correction suites. For this tutorial, we'll use Adobe After Effects version 5.5, but this effect can be easily created using any of the applications mentioned in this book. Additionally, this tutorial can be seen on the companion CD-ROM.

Step One

First, we want to import our clip into After Effects by selecting Import from the File drop-down menu located at the top of the After Effects desktop display. After importing the clip, it should be visible in the Project window.

Step Two

Our next step is to drag and drop the clip into the timeline. When this has been completed, the image will be seen in the After Effects display area.

Step Three

First, we want to add an adjustment layer. This is done by clicking on the After Effects' Layer drop-down menu located at the top of the desktop. Once we've selected Layer, we select New, then Adjustment Layer.

Step Four

Once Step Three has been completed, we see an Adjustment Layer appear on the timeline. Next we double-click on the Adjustment Layer area of the timeline. This will open up a new window.

Step Five

Now, we want to create a mask by right-clicking with the mouse within the display and selecting New Mask. This creates a rectangular mask with yellow lines. There should also be little squares on each corner and at the midpoint of each line.

Step Six

Once Step Five has been completed, we'll change the shape of our mask. To do this, we right-click and select Mask Shape. This will open up another window, and at the bottom of this window are a series of options. We click Oval and confirm this by clicking OK; the control window will close and we'll have our oval mask.

Step Seven

If we now click with our mouse on the oval mask and move our mouse, the entire shape will move. However, we want to actually alter the size of our oval mask. In order to accomplish this, we double-click on one of the points so that we see a black rectangular shape along with our yellow oval shape. If we now click on any of the small yellow squares and move our mouse, that portion of the oval matte will change shape. At this point, let's just decrease the size of our oval mask and proceed to giving it a little rotation.

Step Eight

In the center of the display we see a circle with a cross through it, along with another smaller circle with small points on each side. If we click on the circle with the cross through it, you'll be able to rotate the shape. Clicking on the smaller circle and moving it will alter the rotation axis. For this exercise, let's just rotate the mask a little on the center axis.

Step Nine

We would like to have the correction occur outside this area like a traditional vignette. In order to achieve this, let's right-click on our mask and select Inverse.

Step Ten

Now it's time to put a correction in this area. So, we'll proceed by selecting Adjust from the Effect drop-down menu. Following this, we'll select Brightness & Contrast. Once this has been done, the Effect Controls window will automatically open.

Step Eleven

Next, we'll click on the Brightness control and adjust it by using the slider control. Now our main viewing display will show our picture with a hard-edged vignette around it.

Step Twelve

Now, let's soften up the edge a bit. This is easily done by right-clicking on our mask and clicking Mask Feather. Once this has been done, another adjustment window will open up. If we type in 150, for example, and then click OK, this will close this window and apply the chosen amount of softness to the shape.

Step Thirteen

Now, let's examine what we've done. We should see an oval shape with the normal color on the inside, and it should be darker on the outside, with a nice soft transition. We could also apply some color correction if we choose simply by applying another tool. However, in this situation we just wanted to darken the edges.

Step Fourteen

We've examined what we've created and we're quite pleased, so it's render time once again. We'll simply select the Composition drop-down menu located at the top of the main After Effects desktop, followed by selecting Add To Render Queue. Following this, the Render Queue window will open, and clicking Render starts the process.

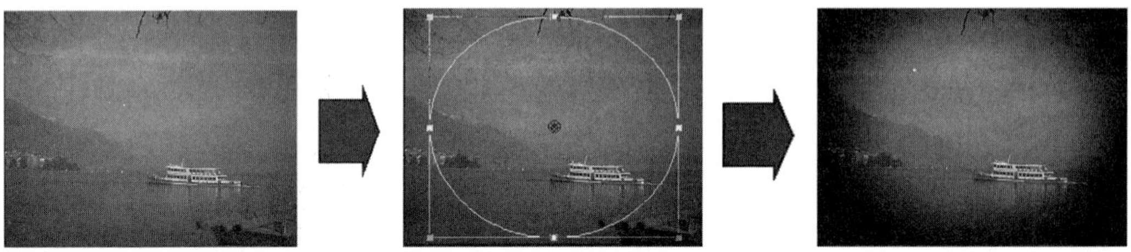

Image 7-16 *Creating a Vignette Effect. (See also Color Plate 7.)*

Making Multiple Color Gradations

In an earlier tutorial, we constructed a color gradation using the plug-in application from Digital Film Tools. In this tutorial, we're going to construct multiple color gradations within the image without the use of a plug-in application. For this tutorial we'll be using Adobe's After Effects version 5.5.

Step One

First, let's import our clip into After Effects by selecting Import from the File drop-down menu located at the top of the After Effects desktop display. After we import the clip, it should be visible in the Project window.

Step Two

The next step is to drag and drop the clip into the timeline. Once this has been done, the image will be seen in the After Effects display area.

Step Three

Now, let's add an adjustment layer. This can done by clicking on the Layer drop-down menu located along the top of the After Effects desktop. Once Layer has been selected, we select New, and then click Adjustment Layer.

Step Four

After completing Step Three, we see Adjustment Layer show up on the timeline. Next, we double-click on the Adjustment Layer area of the timeline. This will open up a new window.

Step Five

Now, we want to create a mask by right-clicking within the display and selecting New Mask. This will create a rectangular mask with yellow lines. There should also be little squares on each corner and at the midpoint of each line. If we double-click on one of the small squares, the yellow frame should turn black with the small squares remaining yellow. In this mode, you can easily adjust the size of the rectangle. For this exercise, let's make it cover only the top end of the image.

Step Six

Now it's time to create our next mask. We do this just as we did the previous one by clicking the Layer drop-down menu, followed by selecting New and then Adjustment Layer. Once this has been done, another adjustment layer should show up on the timeline.

Step Seven

Let's click on our new adjustment layer, which is in the time line. We'll again see the layer display pop up. Next, let's follow up by right-clicking and selecting New Mask.

Step Eight

Once we have completed Step Seven, we'll see a rectangular mask with yellow lines. There should also be little squares on each corner and at the midpoint of each line. If we double-click on one of the small squares, the yellow frame should turn black with the small squares remaining yellow. In this mode we can easily adjust the size of the rectangle. For this exercise, let's make it cover only the bottom part of the image.

Step Nine

We have two masks constructed on two adjustment layers. Now, it's time to put some corrections in there. It's important to note that the active mask is the one on which the adjustable points are visible. If we double-click on Adjustment Layer 1, we will see the mask shape appear. Let's choose a color correction tool from the Effects drop-down menu. For this tutorial, let's choose the color balance tool from the Adjust menu. If we click on any of the controls and adjust the sliders, the color change will take place only in this area.

Step Ten

Following our color correction, we want to soften this mask by left-clicking on the mask, followed by right-clicking and selecting Mask Feather from the pop-up menu. For this exercise, let's type in 200, and then click OK, which will close the window. The resulting softness will then be displayed in the After Effects window.

Step Eleven

Now, we want to double-click on Adjustment Layer 2, and this mask shape will now appear. Let's again choose a color correction tool from the Effects drop-down menu. For this tutorial, let's choose the color balance tool from the Adjust menu. We can then click on any of the controls and adjust the sliders; the color change will take place only in this mask area.

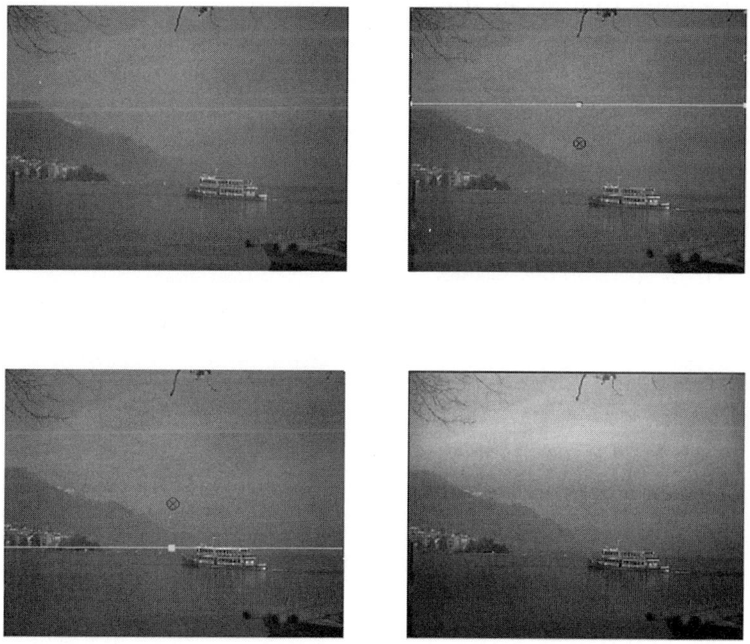

***Image** 7-17 Creating Multiple Gradations.*

Step Twelve

Following our color correction, we want to now soften this mask as well, which we'll do by left-clicking on the mask, followed by right-clicking and selecting Mask Feather from the pop-up menu. For this exercise, we'll type in 200 and then click OK. The window will close, resulting in the softness being displayed.

Step Thirteen

Now, let's close the adjustment layer window and examine the two gradations. If we click in the black area outside the picture area, the mask lines will disappear so that we can see the results.

Step Fourteen

We've examined what we've created and we're very happy with the results, so it's time to render it once again. We simply select the Composition drop-down menu located at the top of the main After Effects desktop display, followed by selecting Add To Render Queue. Following this, the Render Queue window will open, and clicking Render will start the process going.

APPENDIX A

Additional Resources

Anderson, Gary H. *Video Editing and Post-Production: A Professional Guide*. White Plains, NY: Knowledge Industry Publications, 1993.

Bushby, Alex. *A-Z of Film, Television and Video Terms*. Routledge, 1994.

Ebersole, Samuel. *Focal Encyclopedia for Electronic Media*. Newton, MA: Focal Press.

Ishihara Color Vision Test. Developed by Dr. Shinobu Ishihara.

Jones, Stuart Blake, Kallenberger, Richard H., Cvjetnicanin, George D. *Film into Video: A Guide to Merging the Technologies*, 2nd Edition. Newton, MA: Focal Press, 2000.

Katz, Ephraim. *The International Film Encyclopedia*. New York: HarperCollins, 1979.

Konigsberg, Ira. *The Complete Film Dictionary*. New York: Meridian Books, 1987.

Krupnick, Michael A. *Electric Image: Examining Basic TV Technology*. White Plains, NY: Knowledge Industry Publications, 1990.

Parramon, Jose M. *Color Theory*. Watson-Guptill, 1989.

Poynton, Charles. *A Technical Introduction to Digital Video*. New York: Wiley and Sons, 1996.

Schneider, Arthur. *Electronic Post-Production Terms and Concepts*. Newton, MA: Focal Press, 1990.

Watkinson, John. *The Art of Digital Video*. Newton, MA: Focal Press, 1990.

APPENDIX B

World Television Standards

NTSC

- American Samoa
- Antigua
- Antilles
- Aruba
- Azores (U.S. Military)
- Bahamas
- Barbados
- Barbuda
- Belize
- Bermuda
- Bolivia
- Canada
- Cayman Islands
- Chile
- Colombia
- Costa Rica
- Cuba
- Curacao
- Diego Garcia
- Dominica
- Dominican Republic
- Ecuador
- El Salvador
- Fiji
- Galapagos Islands
- Greenland
- Grenada
- Guam
- Guatemala
- Guyana
- Honduras
- Jamaica
- Japan
- Johnston Islands
- Korea, South
- Leeward Islands
- Marshall Islands
- Mexico
- Micronesia
- Midway Islands
- Montserrat
- Myanmar
- Nicaragua
- Northern Mariana Is.
- Okinawa
- Palau
- Panama
- Peru
- Philippines
- Puerto Rico
- Samoa, American
- St. Helena
- St. Kitts/Nevis
- St. Lucia
- St. Vincent
- Suriname
- Taiwan
- Tonga
- Trinidad & Tobago
- United States
- Venezuela
- Vietnam
- Virgin Islands

PAL	Abu Dhabi, Afghanistan, Algeria, Andorra, Angola, Argentina, Ascension, Australia, Austria, Azores, Bahrain, Bangladesh, Belgium, Bhutan, Bosnia-Herzegovina, Botswana, Brunei, Cameroon, Canary Islands, Cape Verde, Channel Islands, China, Christmas Island, Cook Islands, Croatia, Cyprus, Denmark, Dubai, Easter Island, Estonia, Ethiopia, Falkland Islands, Faroe Islands, Finland, Gambia, Gaza/West Bank, Germany, Ghana, Gibraltar, Guinea-Bissau, Holland, Hong Kong, Iceland, India, Indonesia, Ireland, Israel, Italy, Jordan, Kampuchea, Kenya, Kiribati, Kuwait, Laos, Latvia, Lesotho, Liberia, Liechtenstein, Lithuania, Luxembourg, Macao, Macedonia, Madeira, Malawi, Malaysia, Maldives, Malta, Monaco, Montenegro, Mozambique, Namibia, Netherlands, Nepal, New Zealand, Nigeria, Norfolk Island, Norway, Oman, Pakistan, Papua New Guinea, Paraguay, Poland, Portugal, Qatar, Romania, Samoa, Western, San Marino, Sardinia, Sao Tome & Principe, Saudi Arabia, Serbia, Seychelles, Sierra Leone, Singapore, Slovakia, Slovenia, Solomon Is., South Africa, Spain, Sri Lanka, Sudan, Swaziland, Sweden, Switzerland, Syria, Tanzania, Tasmania, Thailand, Tristan da Cunha, Turkey, Uganda, United Arab Emirates, United Kingdom, Uruguay, Vanuatu, Vatican City, Republic of Yemen, Yugoslavia, Zambia, Zimbabwe

SECAM	Armenia	French Polynesia	Martinique	Togo
	Azerbaijan	Gabon	Mauritania	Tunisia
	Botswana,	Georgia	Mauritius	Turkmenistan
	Bulgaria	Gilbert Islands	Mayotte	Ukraine
	Burkina Faso	Greece	Moldavia	Uzbekistan
	Burundi	Guadeloupe	Monaco	Vietnam
	Belarus	Guinea Bissau	Mongolia	Wallis &
	Central African	French Guyana	Morocco	Fortuna
	Republic	Iran	New Caledonia	Zaire
	Chad	Iraq	Niger	(Democratic
	Comoros	Ivory Coast	Reunion	Republic of
	Congo	Kazakhstan	Russia	The Congo)
	Czech Republic	Kyrgyzstan	Rwanda	
	Djibouti	Latvia	Saudi Arabia	
	Equatorial-	Lebanon	Senegal	
	Guinea	Lithuania	Slovakia	
	Estonia	Luxembourg	Tahiti	
	France	Mali	Tajikistan	
PAL-M	Brazil			

APPENDIX C

Color Temperature Chart

(*Note*: These Are Only Approximate Measurements)

Light Source	*Color Temperature*
Candle	1,000 K
Household Lighting	2,500 K
Photoflood Lights	3,000 K
Xenon Light	5,000 K
Daylight	5,000–5,500 K
Overcast Day	7,500–10,000 K
Cloudless Blue Sky	11,000–12,000 K

APPENDIX D

Video Recording Format Chart

Video Format	*Input and Output*	*Audio Track Configuration*	*Maximum Playing Time (minutes)*
1″ Type C	Analog Composite	3 (2 audio and 1 time code)	180
8 mm	Analog Composite	1 Stereo AFM 1 Stereo PCM	180, 240
Betacam®	Analog Component	2	30
Betacam SP®	Analog Component	4	90
Betacam SX®	Digital Component	4 digital	184
Betamax®	Analog Composite	2	300, 215
D-1	Digital or Analog Component	4 digital	90
D-2	Digital or Analog Composite	4 digital	60, 126

Video Format	*Input and Output*	*Audio Track Configuration*	*Maximum Playing Time (minutes)*
D-3	Digital Composite	4 digital	246
D-5	Digital Component	4 digital	124
D-6	Digital Component	12 digital	64
D-7 (DVCPRO)	Digital Component	2 digital	184
DCT®	Digital Component	4 digital	208
Digital Betacam®	Digital Component	4 digital	124
Digital S (D-9)	Digital Component	4 digital	104
DVCAM™	Digital Component	2 digital	184
DV/DVC	Digital Component	2 digital	6.9 hours using a DVCAM 184-minute cassette in LP mode
DVCPRO 50	Digital Component	4 digital	90
Digital 8™	Digital Component	2 digital	90
HDCAM	Digital	4 digital	124
Hi-8®	Y/C	1 stereo AFM 1 stereo PCM	180, 240
S-VHS	Analog Y/C	2 analog	480
U-Matic	Analog Composite	2 analog	75
VHS	Analog Composite	2 analog	320, 600

Glossary

The following is a glossary of various terms mentioned in the previous chapters, along with a selection of terms that the reader may find useful during a session or when dealing with clients. It's important to note that many of these terms are industry-related expressions that have become attached to certain functions.

1:1:
This term is applied to the field-to-frame ratio for progressive scanning.

10-Bit:
This term indicates the presence of 1023 levels in the digital signal.

24p:
Best described as a twenty-four frame per second progressive video format.

¾-inch U-matic:
A composite videocassette system that uses ¾-inch videotape encased within a cassette housing.

16 × 9:
A wide-screen television format measured at 16 inches wide and 9 inches high, which ultimately produces a wide-screen television image.

4:1:1:
This format is used in NTSC DV, DVCAM, and DVCPRO. Using this method, the luminance information is at full resolution; however, the chrominance resolution is half of what it is in 4:2:2.

4:2:0:
This format is used for PAL DV and DVD. As in 4:2:2, there is full-resolution luminance; however, the chrominance only has half-resolution on every other scanning line.

4:2:2:
This is widely used in television postproduction and is recorded on machines such as D1 videotape recorders and Digital Betacam® videorecorders. The terminology stands for full-resolution luminance and only half the amount of chrominance resolution.

4:4:4:
This format is seen generally in high-end postproduction systems and contains full and equal amounts of chrominance and luminance information. The first number represents the luminance channel, the second is the R-Y chrominance channel, and the last is the B-Y chrominance channel.

Academy Standards:
Technical parameters set up by the Academy of Motion Picture Arts and Sciences. For example, Academy Film Leader, Academy Aperture, and so on.

Access Time:
The time required for a data-storage device to locate and retrieve data.

ACE:
This is short for the American Cinema Editors.

Active Picture Area:
The part of a video signal that contains the actual picture information, as opposed to sync or other information. The active picture area is 487 lines for NTSC and 576 lines for PAL. The inactive area is known as blanking.

Additive Colors:
The principal colors of red, green, and blue are known as additive colors. Added together, these colors compose white.

Aliasing:
An effect caused by using a low sampling rate of an image. It causes highly textured areas of the picture to change slowly within the picture. When this effect occurs, accurate reproduction of the image can be very difficult or even impossible.

Alpha Channel:
An image channel that stores information used for image compositing. This information is usually in an 8-bit format; however, there are also some 16-bit applications. Generally, only formats such as the PICT or QuickTime™ Animation codec support this type of image channel.

Ambient Light:
This is essentially the light emanating from skylights, windows, room lamps, and the like. In essence, this is the room's normal lighting, rather than lighting that has been set up especially for a particular production.

Amplitude:
The level of an electrical signal. The value is generally measured in values such as volts, millivolts, or IRE units.

Analog:
This term refers to the representation of data by continuous voltage variations. An analog signal can be generated in a component or composite format.

Analog-to-Digital Converter:
Also referred to as an A/D converter, this is an electronic converter that changes analog information into digital information.

Analog Video:
This term refers to the representation of video data by continuous voltage variations. An analog signal can be generated in a component or composite format.

Anamorphic:
An anamorphic system photographs the image with a horizontal squeeze, and then during reproduction unsqueezes the image in order to produce a wide-screen image. In video, a similar method is used for 16 × 9 reproduction. In motion picture theaters, this term is applied to formats such as CinemaScope® or Panavision®, which utilize this process by placing a special anamorphic lens in front of the camera during the shoot, and placing another lens in front of the projector during projection to unsqueeze the image.

Anti-aliasing:
The method of smoothing "steppy" edges created by the aliasing effect.

Application:
A computer software program that performs a specific job or fills a specific need.

Archive:
A long-term storage format for information. Although the term "archiving" generally refers to storage on a CD, tape, or disk, it is also used in the film industry for long-term storage of film material for preservation or stock-footage uses.

Area Isolation:
See Selective Color Correction.

Artifact:
An item that has been added to the video signal as a result of an error in signal processing. Artifacts can take on different forms, depending on whether they are in a digital or analog format.

Artificial Light:
Man-made illumination coming from lighting sources such as tungsten lamps, fluorescent tube lighting, and the like. In most cases the color temperature is lower than that of daylight; however, there are some artificial light sources specially designed to have higher color temperatures, in order to give the effect of daylight color.

ASA:
A rating or measurement that shows the film sensitivity. Originally instigated by the American National Standards Institute.

ASC:
This stands for the American Society of Cinematographers, which was established in 1919 and offers membership only by invitation.

Aspect Ratio:
This is the ratio of the image's height to its width. There are several ways this can be labeled; for instance, a standard 4 × 3 aspect ration picture may also be shown as 1.33 : 1. The latter is generally used more for theatrical film applications. For instance, a CinemaScope® film is measured with an aspect ratio of 2.35 : 1.

Assemble Edit:
A video edit that adds information to the end of a recording; it essentially continues the control track on the tape from the point at which the recording ceased.

Assistant Cameraman:
The member of the crew responsible for maintaining the paperwork, loading the camera, pulling focus, and so on.

Assistant Director:
The assistant director removes some of the load from the director's shoulders in order to allow him or her more creative freedom on the set.

ATM:
Also known as Asynchronous Transfer Mode, essentially a network transfer protocol for data in small units known as cells or packets.

Available Light:
The light that may be present in a room or location during a shoot without the use of any other additional lighting, photofloods, and so forth.

Background Processing:
A function that allows the computer to perform certain operations while another activity is in process, such as printing while word processing is performed.

Back Light:
A lighting source that emanates from behind the subject. It usually results in a very soft rendition of the subject's detail, as the brighter background tends to create a silhouette effect.

Back Lot:
The exterior area in a film studio that has special sets constructed for doing the exterior shots of a production without going on location.

Backup:
A duplicate file created in case a problem arises during work with a particular application. Backups are frequently made when no problems have occurred, simply to protect the information in the event that the original file is damaged or destroyed.

Bandwidth:
A range of frequencies in an electronic system, whereby the larger the bandwidth the better the performance.

Best Light:
This term pertains to a color correction that is somewhere between a scene-by-scene color correction and a one light color correction. The levels are balanced in order to facilitate a tape-to-tape color correction at a later time.

BETACAM®:
A videorecording format developed by the Sony Corporation. The information is recorded onto a half-inch (12.7-mm)-wide videotape enclosed within a cassette. The format has a maximum recording time of thirty minutes.

Betacam SP®:
The Sony Corporation released the Betacam SP® format in 1987. The format supports videocassettes with a longer running time than Betacam. As with its Betacam counterpart, Betacam SP uses a half-inch (12.7-mm)-wide videotape enclosed within a cassette.

Betamax®:
A videocassette format originally developed by the Sony Corporation and intended for consumer use. It utilizes a half-inch magnetic videotape enclosed within a cassette.

Bit:
A single fragment of data in the digital signal, which could be either a 1 or 0.

Black and Code:
This term generally refers to tape stock that has been prerecorded with black and time code so that program material may be inserted over it.

Black a Tape:
This term refers to recording a black burst signal for the duration of a tape. This is basically the same as when an operator "black and codes" a tape for later insert editing of program material.

Black Stretch:
A function that increases the range of dark information in the picture.

Bleeding:
This term generally refers to a video artifact that shows as a blurring or bleeding of the chroma information. This is seen frequently when viewing composite analog video material, and also when viewing analog home video formats.

Booking:
Reserving time in a video facility for a session. Generally one booking refers to one session.

Brightness:
This term pertains to the luminance level for an image. Its levels are generally within the range of 0 to 255, with 0 being black and 255 being white.

Broadcast:
This term refers to information to be used for television transmission.

Buffer:
This term generally refers to a RAM storage area for data information. Its general purpose is to be used by the computer's CPU to work with data or information prior to output.

B-Y R-Y:
The "Y" stands for the luminance information, and the "B" stands for blue and the "R" for red.

Byte:
A group of eight memory bits.

Camera Report:
A daily report that has information regarding the day's shoot.

Candela:
The illumination capability from a particular type of light. It is also referred to as "cd" units.

Capture:
The process of converting video information into a digital form for use within a computer. This is generally used when working with nonlinear applications. It is also referred to as digitizing.

CCD:
This is short for charge-coupled device, a photosensing unit that is used in some telecines and television cameras.

CCIR (Comité Consultatif International des Radiocommunications):
This is a committee based in France that sets up and approves audio and video standards.

CD:
The abbreviation for the term "compact disc." There are several different formats, such as CD-R and CD-RW, which are used for data storage.

Channel:
This is one portion of the image's information. Generally an image contains a red, green, and blue channel.

Chroma:
This term pertains to the image color and is composed of saturation and hue.

Chrominance:
This term pertains to the color information within composite video.

CinemaScope®:
A copyrighted name for a 2.35 : 1 film format containing an anamorphic squeeze, which was started by the 20th Century Fox film corporation in 1952. It is actually derived from the format Hypogonar, which was invented by Professor Henri Chrétien in the 1920s. The image is anamorphically squeezed in the camera, then unsqueezed with a special lens during projection, which results in a wide-screen image.

Clapperboard:
A board or slate used to identify different takes of a production. It also has a hinged stick that generates an audible clap, which is used to aid in the synchronizing of picture and sound.

Client:
A customer who brings his or her material to the postproduction facility.

Clip:
This term pertains to (1) a finite portion of source information, or (2) one particular shot within the timeline.

Clipped Whites:
This term pertains to luminance information that has been adjusted so high that the levels have been limited, resulting in a loss of detail.

Clipping:
A method of limiting the luminance information to ensure that the levels are clipped, or cut off, within legal limits. When levels exceed the maximum, they are clipped, resulting in a loss of detail.

Color Balance:
The adjustment of color in order to arrive at a proper color rendition of the image.

CMY:
This term stands for the three subtractive primary colors: cyan, magenta, and yellow.

CMYK:
A color-encoding process that uses the subtractive primary colors along with black, to produce color with a printer device. The term CMYK literally stands for cyan, magenta, yellow, and keyline, or black.

Color Bars:
A test pattern composed of eight bars of color, which is an industry standard for adjustment and for ensuring continuity throughout the postproduction process.

Color Burst:
A signal used for synchronizing the image's color information.

Color Contamination:
A bleeding between colors, specifically colors within an object that should be purely white or black.

Color Correction:
The method of selecting the correct balance of color and image density to achieve the proper color rendition, or to simply create a specific look.

Color-Correction Computer:
The color-correction computer allows the operator or colorist to perform color correction and to memorize the changes made on a scene-by-scene basis.

Color Depth:
The available color range within the image, such as 8 bit or 16 bit.

Color Difference:
This is also referred to as the color difference signals. They subtract the luminance information, creating R-Y and B-Y, with the luminance information being Y.

Color Frame:
The video fields needed to make up the field, frame, and color subcarrier synchronizing. For example, in NTSC the sequence is four fields, and in PAL, eight fields.

Color Gamut:
The range of colors available for a system. For example, there are certain colors that cannot be displayed accurately on certain systems.

Color Grading:
See Color Correction.

Color Picture Monitor:
This is essentially a television monitor, similar to a commercial television.

Color Reference Burst:
A signal used within composite video for the purpose of synchronizing the color information and establishing the hue.

Color Temperature:
A measurement for a light source's color by degrees on a scale known as Kelvin. The higher the temperature, the cooler the color of the light. A lower color temperature displays a warmer lighting color.

- Candle Light = 1000 K
- Photoflood Light = 3000 K
- Household Lighting = 2500 K
- Daylight = 5000–5500 K

Color Timing:
The method of selecting the correct balance of color and image density in a motion picture film laboratory to achieve the proper color rendition. The operator is known as a color timer.

Color Wheel:
A circular display that shows the color hues available. It is used widely in color correction tools for primary and secondary color correction.

Component Video:
The component video system is composed by using three separate signals of red, green, and blue, or RGB components. Another variation is the Y, R-Y, B-Y method used by some manufacturers. In the variation of Y, R-Y, B-Y, "R" stands for red and "B" for blue. The "Y" pertains to the luminance information, and the green information is generated from the other components.

Composite Video:
The system of composite video is created when the three separate red, green, and blue signals, along with the luminance and sync information, are combined through the use of a composite video encoder, which creates a signal in NTSC, PAL, or SECAM. Although a certain amount of quality is lost, it allows for greater ease during transmission.

Compositing:
A method of combining layers within a compositing application—for instance, the addition of titles or key effects.

Compression:
Removing unimportant information in order to reduce file size.

Compression Ratio:
The ratio of the compressed to the uncompressed picture.

Continuous Tone:
Constant brightness exhibited throughout a picture.

Contouring:
An effect that is like banding, caused by low resolution in the highlights of the image.

Contrast:
The range from dark to light information seen in a video image.

CPU:
The central processing unit, which is the information-processing chip within a computer.

Cropping Tool:
A tool for cropping, or cutting off, portions of the image.

CRT:
This is the abbreviation for cathode-ray tube.

Crushed Blacks:
This term refers to the loss in details in the dark areas of an image due to excessive compression of the black areas of the image.

Curves Graph:
The curves graph shows the lift information at the lower-left portion of the graph, and the gain or luminance information at the top-right portion of the graph. The information between these two areas on the graph represents the midtones.

Cutaway Shot:
This type of shot can be used for various purposes. It is generally intended to extend the viewer's angle of view within the scene by cutting away from the principal action.

D-1:
The D-1 component digital format was introduced in 1986 and has since become the industry standard for noncompressed digital videorecording. The recording is

made on a ¾-inch-wide videotape enclosed within a cassette. The maximum recording time for the format is 90 minutes.

D-2:
The D-2 format is a composite digital format that records information onto a ¾-inch-wide videotape enclosed within a cassette and runs at a speed of 5.185 inches per second, or 131.7 millimeters per second.

D-3:
The D-3 format is a composite digital format developed by the Japanese Broadcasting Company NHK in 1991. It records information onto a half-inch-wide videotape enclosed within a cassette, and runs at a speed of 3.28 inches per second, or 83.88 millimeters per second.

D-5:
The D-5 component digital format was released in 1994 by the Panasonic corporation. The format uses the same type of cassette as the D-3 format. The format also has four digital audio channels and a dedicated time-code channel.

D-6:
D-6 is a compression-free component digital HDTV recording format developed by Philips and Toshiba. The recording is made onto a ¾-inch-wide videotape enclosed within a cassette, providing a maximum recording time of 64 minutes.

D-7:
The D-7 format is also known as DVCPRO and was developed by the Panasonic Corporation. The format was developed for industrial and broadcast uses, generally intended for electronic news-gathering applications. It records the information using a special MP, or metal particle, 6-mm videotape with a maximum recording time of 184 minutes.

D-9:
The D-9 format, also known as Digital S, was developed by the JVC Corporation and is a component digital format. It has a sampling rate of 4:2:2, and the recording is made on a half-inch metal particle videotape enclosed within a cassette. It actually uses the same cassette shell that is used for the VHS format.

Daily:
This term generally refers to a one-light motion picture film print or videotape film transfer that has been made of newly photographed material. Dailies are generally made in order to check that all parameters of the shoot are correct.

Data:
Information within the computer in a digital format.

DCT:
The DCT format is a component digital format. The term DCT stands for Digital Component Technology®, which was developed by the Ampex Corporation. The format records onto a ¾-inch- or 19-mm-wide videotape enclosed within a cassette. The format has a compression ratio of two to one.

DDR:
This device, also known as a digital disk recorder, can record audio and video onto an internal disk.

Default Setting:
This is generally a factory-preset value within an application. If the user does not change an option, it is at the default setting.

Degauss:
The method of erasing the information contained on a magnetic tape by using a special device known as a degausser.

Densitometer:
A device that measures the degree of light transmitted by the subject.

Depth of Field:
A terminology referring to distances viewed in focus by a camera. For instance, wider f-stops will decrease the depth of field, and smaller f-stops will increase it, thus increasing the range of objects viewed in focus.

Detail:
The visibility of minuscule information within the image.

Digital:
Digital video consists of the analog values represented in the form of binary numbers. When the analog values are measured, this is known as sampling. The higher the frequency of sampling, the better the image detail within the signal. Analog signals can be converted to a digital format with an analog-to-digital converter. Alternately, the reverse process can be carried out by using a digital-to-analog converter, or a D/A.

Digital Betacam®:
Digital Betacam uses a half-inch (12.7 mm) videotape enclosed within a cassette that is specially marked for use within a Digital Betacam machine. It is a component digital format with a two-to-one compression ratio, along with four digital audio tracks and a time-code track.

Digital Disk Recorder:
A device, also known as a DDR, that can record audio and video onto an internal disk. Digital disk recorders come in many formats, some of which are only capable

of recording a short duration of video, while others can record many hours of information.

Digital Recorder:
A recording device that records video and audio in a digital form.

DIN:
Stands for Deutsche Industrie Norm, a rating system for film speed.

Direct Digital Interface:
The method of connecting several devices together without going through a D/A converter.

Direct Lighting:
This type of lighting is usually quite hard, and is generally seen in electronic news-gathering applications when there is a single light source illuminating the subject.

Director of Photography:
The person who has the responsibility for photographing the program content. The director of photography, or DP, is responsible for ensuring that all parameters of the shoot are correct.

Downtime:
Downtime occurs when there is a system failure within the suite. When downtime occurs during a session, it is usually deducted from the working time.

DP:
See Director of Photography.

Drag and Drop:
The method of moving information to different places within an application by dragging it with the mouse and dropping it into the desired location.

Drop-Frame Time Code:
This is a time-code format found in NTSC, which runs at thirty frames per second. It deletes two frames every minute and suspends the process every tenth minute.

DUB:
A duplicate, generally made from another videotape.

DV:
The DV format, also referred to as DVC, is a component digital recording format that has a five-to-one compression ratio. It also supports up to two channels of digital audio. The information is recorded upon a special 6-mm metal-evaporated videotape enclosed within a cassette.

DVCAM®:
The DVCAM format from Sony is a component digital recording format. It supports up to two channels of digital audio and has a maximum recording time of 184 minutes, recorded on a special 6-mm metal-evaporated videotape enclosed within a cassette. The format is generally intended for professional or industrial applications. In NTSC, it records with a sampling rate of 4 : 1 : 1, which means that the luminance information is at full resolution; however, the chrominance resolution is half of what it is in 4:2:2. When used in PAL, the sampling rate is 4:2:0, which means that there is full-luminance resolution, as in 4:2:2, but that the chrominance only has half-resolution on every other scanning line.

DVCPRO:
See D-7.

DVD:
The DVD disc provides a higher amount of storage space than the conventional CD-ROM. For home video applications, DVD uses a digital sampling of 4:2:0, which means that there is full-resolution luminance; however, the chrominance only has half-resolution on every other scanning line.

DVE:
This stands for digital video effects, and refers to devices intended for digital manipulation of the image. These devices can perform a variety of functions, including pan, tilt, zoom, and even wipes.

EBU:
The abbreviation for European Broadcasting Union.

EBU Color Bars:
This is a color bar test signal developed by the European Broadcasting Union, which uses 100 IRE luminance.

EBU Time Code:
This is a time-code system developed by the European Broadcasting Union (EBU) and is a 25 frame per second count due to the characteristics of the PAL standard. It displays the count in hours, minutes, seconds, and frames.

Edit Decision List:
Otherwise known as an EDL, this is a edit list generated by an electronic editing system. It generally contains lists of the in points and out points of the source and record machines.

Edit Point:
The location at which an edit is created, whether it be a physical edit done on motion picture film or an electronic edit in the video domain.

EDL:
Also known as an edit decision list, this is an edit list generated by an electronic editing system. It generally contains lists of the in points and out points of the source and record machines.

EIA:
The Electronic Industries Association, an association that produces a catalog pertaining to television and audio standards.

EIA Color Bars:
A color bar chart developed by the Electronic Industries Association, which displays seven color bars with 100 percent chroma saturation and 75 percent amplitude.

Encoder:
A unit that combines the channels of a component signal in order to create a composite signal.

Engineering Setup:
This can be an allotted time for the engineers within a facility to set up a particular piece of equipment. It also refers to a special setup area within a software application that is not intended to be adjusted by the operator.

EPS:
Encapsulated PostScript, a graphic format.

Erasure:
The removal of the material recorded upon the existing tape.

ESU:
Also known as engineering setup, which is generally an allotted time for the engineers within a facility to set up a particular piece of equipment. It also refers to a special setup area within a software application that is not intended to be adjusted by the operator.

Ethernet:
A method for transmitting information between computers. It can transmit at the speed of around 10 million bits per second.

European Broadcasting Union:
An organization based in Europe that sets up various broadcast standards.

Event:
Events are the settings programmed into a color-correction device, which generally pertain to either an existing scene or a frame.

Event List:
A list of events in a color-correction device that pertain to either a particular session or a film.

Export:
A method of moving information from one application to another, in such a way that the receiving application will be able to read it.

Exposure Meter:
A device used for measuring an exposure.

Facility:
A company or business that offers the services of postproduction.

Fade In:
This term refers to the gradual adjustment of the luminance from black to normal level. This is usually done at the beginning of a scene.

Fade Out:
Lowering the video from normal level to black. This is generally done at the end of a scene.

FCC:
The Federal Communications Commission, which is the U.S. commission that looks after television transmission within the country.

Fiber Optics:
A light-transmitting method that utilizes glass or plastic fibers.

Field:
Each video frame in an interlace system is composed of two fields, one with even-numbered lines and the other with the odd-numbered lines.

Field Frequency:
This term refers to the quantity of fields. For example, PAL contains 50 fields per second, and NTSC contains 60 fields per second.

Field Store/Frame Store:
A recording device that stores still frame images.

File:
A named set of information, on a computer, that can be named, changed, saved, sent, deleted, and so on. A file can contain various types of data, for example text or images.

File Extension:
The last section of a computer filename, such as .tif or .bmp.

File Format:
A particular type of file, for instance a GIF, TIFF, or DOC file.

Fill Light:
The fill light actually fills in the shadows created by the main key light in the scene. Using this light will decrease the scene's contrast.

FireWire®:
An external bus developed by Apple Computer, Inc., which can transfer data at around 400 Mbps.

Flat:
This term generally refers to an image without contrast.

Flat Lighting:
A soft-lighting technique that results in minimum subject contour.

Footage Counter:
A device that measures film footage.

Format:
There are many formats used to record either digital or analog information. Certain recording formats are intended for either consumer or professional applications.

Formatting:
The method of formatting a videotape by recording a color bar signal along with an audio tone followed by black and then the program information.

fps:
Frames per second.

Frame:
Either a single film frame or a video frame. In the case of a video frame, it is composed of two fields.

Frame Grab:
The process of "grabbing," or storing, a particular video frame.

Frame Rate Conversion:
The method of converting the frame rate of the material to a different frame rate.

Frame Store:
A device that stores frames for later reference, generally used for color matching.

Front Lighting:
This is actually the most basic type of lighting, which hits the subject head-on. Although the subject may be well lit, this type of lighting is usually quite flat.

F-Stop:
A lens adjustment that determines the opening of the iris within the lens.

Full-Motion Video:
Moving video material viewed on a computer display.

Gain:
Refers to the adjustment of the image's luminance information.

Gamma:
The adjustment of the contrast between the light and dark areas of the picture.

Gamma Correction:
Adjustment values applied by the Gamma control.

Gamma Curve:
A graphic view of the current gamma value. The upper-right corner of the graph or display will show the white information, and the lower-left, the black information. In between these two areas is the gamma information.

Gauge:
Generally refers to the motion picture film's width, such as 16 mm or 35 mm.

General Purpose Interface:
Also referred to as GPI, this is a method for a computer to control or trigger functions of another piece of equipment.

GIF:
Graphics Interchange Format, a graphics file format introduced by the firm CompuServe.

Gigabyte:
A form of data measurement, otherwise referred to as GB, equal to around one thousand million bytes.

Global:
A type of correction that is carried out over a series of events. Generally this is done to apply an overall trim to an existing correction.

GPI:
See General Purpose Interface.

Grains:
The grains within the emulsion of film stock.

Gray Scale:
A range of shades starting at white and continuing through black.

GUI:
The abbreviation GUI stands for the term "graphical user interface." It pertains to the control interface used by the operator for controlling the various options of a particular application.

Handles:
Additional material at the start and end of the material to be used in the edit. The use of handles is especially necessary when dissolves are to be added during the compositing session.

Hard Disk:
A large-capacity data-storage disk, generally mounted within a computer. There are also removable hard disks that enable transporting large amounts of data from one computer to another.

Hard Record:
The point at which recording starts on a blank tape. Generally, the formatting of the videotape starts with the hard record, and from then on an electronic edit will be used.

Hardware:
All the physical parts of a computer, for instance the monitor, hard disk, and so on.

HDTV:
High-definition television. HDTV generally refers to the various ATSC standards.

Head:
Used to identify the start of either a film roll or videotape.

Helical Scan:
Diagonal tracks recorded on the videotape.

Hertz:
Unit of measurement equal to one cycle per second.

Hi-8:
The Hi-8 format is an analog format that can receive a video input in composite or Y/C mode, which separates the chrominance and luminance information. It was introduced by Sony in 1989. It records the information onto an 8-mm-wide metal-evaporated videotape enclosed in a cassette. Basically a consumer format, it is sometimes used for professional applications also.

High-Definition Video:
A video standard that has a high resolution.

Histogram:
The histogram display gives the operator the ability to view the image's dynamic range and contrast properties. The display is broken down into an X and a Y axis, where the X defines the pixel's luminance, and the Y shows the quantity of pixels at a level of luminance. The histogram has a measurement scale from 0 to 255, with 0 being black and 255 being white.

Holdout Matte:
Generally used for compositing, this is a high contrast image that is used to isolate parts of the image.

Horizontal Resolution:
Generally refers to the definition of the signal, with more lines providing better resolution. For instance, Betacam has better resolution than VHS.

Hue:
This term refers to the particular color that is being selected, and essentially pertains to the range of available colors.

I-Box:
A format that displays either a wide-screen or anamorphic image's entire content on a television screen. The upper and lower portions of the television image are usually replaced with black information. This format is also referred to as letterbox.

IEEE 1394:
This is the standard for FireWire, which was created by Apple Computer, Inc. FireWire is an external bus that can transfer data at around 400 Mbps.

Illegal Video:
This term refers to signal levels that do not fall within the range of a particular broadcast standard—generally either chroma levels that are excessively high, or luminance levels that exceed 100 IRE.

Image:
The subject material captured by the camera.

Image Area:
The area of the film that contains the image to be viewed.

Image Processing:
The process of loading an image into a system with the intent of performing a correction or manipulation.

Image Resolution:
The level of detail in either a photographic or video image.

Image Size:
The size of an image, whether it be a photograph or an image viewed on a video monitor.

Import:
To bring information into an application from another application or storage device.

Insert Edit:
An edit mode in which the audio or video is inserted or recorded over an existing control track on the videotape.

Interlace:
The process in which one picture is composed of two fields, with the first field containing odd line numbers and the second field having even line numbers.

IRE:
A unit of measurement named for the Institute of Radio Engineers. One IRE is equal to 1/140 of the peak-to-peak voltage in a composite video signal.

Iris:
An adjustable diaphragm within a camera lens. Its purpose is to adjust the amount of light that passes through.

ISO:
An isolation reel, recorded from a designated camera during a production. It is also short for the International Organization for Standardization.

ITU-R 601:
A standard set by the International Telecommunications Union for standard definition component digital video.

JPEG Compression:
A compression method using a lossy compression algorithm. JPEG is short for Joint Photographic Experts Group.

Kelvin:
The Kelvin scale measures the color of light.

Keyframes:
The reference frames within an effect.

Key Light:
The main light on a set, which illuminates the prime subject of the scene. It's the cinematographer's tool for riveting the viewer to the prime area of the image.

Kilobyte:
An amount of data that is 1024 bytes in size.

Kilohertz:
A measure equal to one thousand cycles per second. Abbreviated kHz.

Knee:
A lighting effect also referred to as a soft clip.

Lag:
An image blurring caused by a CRT or tube video camera.

Leader:
The portion of either a film or videotape that starts off the roll. In the case of film, it is generally a countdown or clock.

Letterbox:
A format that displays either a wide-screen or anamorphic image's entire content on a television screen. The upper and lower portions of the television image are usually replaced with black information. Also known as I-Box.

Level:
Refers to degree of adjustment for a particular function. For example, adjusting the level of luminance will adjust the overall luminance properties of the image.

Lift:
This control, which is also referred to as pedestal, adjusts the image's black level.

Linear Editing:
This term refers to standard electronic editing in which there are source and record tapes, and the system has to shuttle to the areas that are being used.

Log:
A list of data regarding a particular program, which might have the take, scene numbers, defects observed, and so on.

Long Shot:
A wide-angle, distant shot captured by the camera. It is generally used to establish the location.

Lossless Compression:
A compression method that does not result in any discarding of information.

Lossy Compression:
A method of compression that discards data.

LTC:
Longitudinal Time Code, a time-code signal that is generally in the form of an audio signal.

Luma:
An abbreviation of luminance.

Lumen:
The amount of illumination that emanates from the source.

Luminance:
The level of brightness in the image.

Luminance Channel:
A channel that shows the image's brightness.

Macro:
In computer applications, a chain of commands that can be activated by simply pressing one key.

Maintenance Engineer:
The technician who repairs and installs equipment within a postproduction facility.

Mark In:
The selection of the relative position of the material to begin an edit.

Mark Out:
The selection of the relative position of the material to end an edit.

Mask:
A method of limiting, or masking off, an area of an image. This is generally done in order to have a particular correction occur only in a selected area of the image. Masks can be created using different methods: they can be generated by using a key signal, or even manually drawn.

Master:
The finished videotape that has been completed within the postproduction process.

Master Clip:
The main clip that originates from the master source material.

Mastergrade:
This is also referred to as tape-to-tape color correction, and is generally a fine-tuning of the color correction that was done during the initial telecine transfer.

Master Shot:
Also known as a wide shot or establishing shot. It establishes the location for the viewer.

Master/Slave:
This term refers to the master VTR's control of multiple decks by means of a slave function.

Matte:
A signal that cuts a piece from the selected picture and allows replacement with another image or signal.

Matte Key:
An effect that permits one section of an image to be keyed on top of another.

Medium Shot:
A camera angle that is a little closer than a long shot. It generally follows a set of close-up shots.

Megabyte:
Approximately 1,000,000 bytes of data. Abbreviated as MB.

Megahertz:
One million cycles per second. Abbreviated as MHz.

Menu Display:
A display located within the application desktop. It provides control over specific features of the application.

M-JPEG:
Motion JPEG, a video file containing JPEG compression. This form of compression is used frequently with nonlinear editing applications.

Moiré:
An effect produced by lines within the image that counteract with the scanning lines.

Monitor:
A video display device similar to a conventional television, but generally without a tuner, because it is meant only to accept a direct connection with equipment within the facility.

Monochrome:
A term that essentially refers to black-and-white material.

MOS:
This term is short for "mit out sound." It is said to have originated from a Hollywood-based German director who would use this expression to describe film takes that were without a sound track.

Moviola:
A film-editing machine incorporating a viewing screen that allows the editor to view the image and sound track. The early models were upright versions, and later ones, which were known as flatbed models, contained a larger viewing screen with the film and soundtracks lying horizontally on platters.

MPEG:
This is short for Moving Picture Experts Group, and refers to a series of compression standards.

Multicamera:
This term refers to photographing a scene with multiple cameras. Usually done during a television production, so that multiple angles can be photographed simultaneously, without having to stop. In film productions, the method is used when there is only one chance of catching a particular stunt or effect.

Multimedia:
This term refers to the creation of games, presentations, and the like, with the use of computer systems to combine video and audio effects.

Natural Light:
Generally refers to daylight illumination.

ND Filter:
A neutral density filter, which results in a reduction in light while keeping the subject's color unchanged.

Noise:
Video artifacts observed as snow or grain-like particles.

Non-Drop-Frame Timecode:
This refers to continuous even-number-counting time code, which runs at 30 fps.

Noninterlaced Video:
A technique in which each video frame is actually one field of full resolution, whereas interlace scanning means that one video frame is made up of two fields. Noninterlaced video is also referred to as progressive scan video.

Nonlinear Editing:
A form of editing using a nonlinear system. Generally the material is loaded onto a storage device that can be immediately accessed with the computer running the nonlinear editing application, allowing random access.

NTSC:
The NTSC standard, named for the National Television Standards Committee, is used in the United States, Canada, Japan, and some parts of South America. It's a 525-scan-line interlace system that runs at 60 fields per second.

Offline Editing:
Electronic editing that is done to produce a rough cut of the master that is to be created. Once completed, the offline system generates an EDL, or edit decision list, which will be used during the online edit session to create the finished master.

One Light:
This is generally a single color correction that is applied throughout the material. Because no scene-by-scene correction has been carried out, it is generally very cost-effective.

Online Editing:
An electronic edit session that is performed to produce a finished product, using an edit list produced during an offline electronic editing session.

Operator Control Panel:
Generally a remote-control panel designed to allow the operator access to the unit's controls.

Optical Disc:
A recording disc that records optical information.

Opticals:
This term generally refers to motion picture film optical effects. These are effects created with an optical printer and later incorporated within the original negative.

Oscilloscope:
A testing device that shows the electronic signal on a CRT. It is generally used in the laboratory for testing and repairing electronic equipment.

Out of Sync:
A phrase used when the audio lip sync is out of synchronization with the film being viewed.

Outtake:
A take from a production that has either been cut out of the finished material or excluded due to a defect.

Overexposure:
A bleached image caused by photographing the subject with excessive light.

Overlap:
An overlap generally occurs when there are two videotape rolls, and the second reel has a repeated portion from the end of the first reel.

Overscan:
The mode as seen on consumer televisions, which have a slight portion on the edge of the image masked off. This is unlike underscan, which displays the entire image.

PAL:
The PAL standard is used through most of Europe and Australia. It is a 625-scan-line interlace system. PAL stands for Phase Alternate Line, because the standard utilizes a phase alternate line technique during the encoding process, to achieve

superior color rendition during transmission. The standard is a dual-field interlace format that runs at 50 fields per second.

PAL Plus:
This is essentially a PAL signal that has a 16 × 9 anamorphic squeeze and can be viewed normally on a 16 × 9 wide-screen television.

Pan and Scan:
A method for fitting a wide-screen movie into a 4 × 3 television aspect ratio. Generally, the operator chooses the portion of the frame that will be visible and then programs the image repositioning into a color-correction computer, in order to maintain the important information on the screen. These moves, or pans, are stored in the color-correction computer's event list and then viewed when the material is being recorded.

Panchromatic:
A black-and-white film stock that is sensitive to all colors.

Parade Display:
An alternate display that can be shown on a waveform monitor is known as an RGB, or parade, display. The parade display shows the red, green, and blue channels individually and in sequence. The use of this form of display is the quickest way of viewing the status of each channel's level.

Pedestal:
The reference point in the video signal for the image's black level.

Phase Adjust:
The method for adjusting the color subcarrier's phase. The visual effect is a change in the image's hue.

Phosphor:
The properties, or dots, in the cathode-ray tube that catch the information from the colored electron guns.

PICT:
A graphics file format that comes in two forms, either PICT I or PICT II.

Pixel:
The smallest part of the displayed video image.

Pluge:
This term stands for Picture Line Up Generation Equipment, and is a special test pattern developed by the BBC.

Positive:
An image that is the reverse of a negative. Generally referred to either as a photographic print or a motion picture film print.

Postproduction:
A process that immediately follows the production process and includes the compilation and completion of the production material.

POV:
This term stands for "point of view," and basically means that the camera sees what the subject is looking at.

Preview:
The action of previewing an edit or function prior to finally performing the edit.

Primary Color Correction:
Primary color correction is the method of adjusting the overall color balance within a particular scene. A primary color correction consists of overall adjustments for the lift, gamma, and gain levels, plus color balance controls for each parameter.

Primary Colors:
Generally refers to video's primary colors, which are red, green, and blue.

Process Shot:
A shot that has been superimposed over another background by using a method called rear projection.

Production:
The method of producing a program. Production is followed by the postproduction process.

Programmable:
Controls that can be stored in the application are programmable.

Progressive Scan:
With progressive scanning, each video frame is actually one field at full resolution, whereas interlace scanning means that one video frame is made up of two fields.

Pseudo–Time Code:
This is a form of time code that is generated from tach pulses or other sources, when there is no time code to utilize.

Puality Control:
Also known as QC, this is the method of program inspection intended to ensure that the material is free from problems.

RAM:
The abbreviation of "random access memory," which is unsaved computer memory

Real-Time Effects:
Effects applied to the image that can be viewed in real time, without the necessity of rendering.

Rebook:
To reschedule a session when the unforeseen occurs, such as sickness, equipment failure, or even delayed shipments of materials.

Rec. 601:
Also referred to as CCIR 601 or ITU-R BT.601, this is the digital video sampling standard.

Rec. 709:
This is the standard for HDTV, also known as ITU-R BT.709.

Record Lockout:
A safety device that inhibits the recording device from recording. It can be a switch on the actual recorder itself, or if the record tab on the videocassette in engaged, this will also lock out the recording feature.

Record Monitor:
A dedicated monitor that displays the information being recorded with the videotape-recording machine.

Reference:
A signal that is used for synchronizing other equipment within the facility.

Reference Black:
A reference point that signifies black. For instance, when viewed on a waveform monitor, black is set at 7.5 IRE in NTSC and 0 IRE in PAL.

Register:
To line up multiple images in a sequence.

Release Print:
A positive motion picture film print created for theatrical exhibition.

Rendering:
The method of integrating the viewing material with the particular functions being carried out.

Resolution:
The detail of either a photographic or video image.

Retake:
To reshoot a scene because a problem has been observed.

RF:
The abbreviation for "radio frequency."

RGB:
This refers to the three primary colors: red, green, and blue.

RS-232/RS-422:
A method of communication for controlling video equipment such as VTRs, noise reducers, and the like.

Rushes:
This term can mean either a one-light negative film transfer to video, or a positive motion picture film print that is being projected to allow the production company to view the material that has been shot.

Safe Action Area:
A defined area within which the action for a scene should take place. Information outside this area may be cut off when the scene is viewed on commercial television.

Safe Color Limiting:
The method for bringing the color information for a program into legal or transmittable limits.

Safe Title Area:
A defined area in which the titles for a program should fit, so that they will be properly displayed on commercial television.

Sampling Rate:
The frequency with which the analog values are measured (sampled). The higher the frequency of sampling, the better the image detail within the signal.

Saturation:
The level of chroma information within a particular image.

Scan lines:
Electronic beam scans that move across the cathode-ray tube.

SCART:
A connector that is generally seen in Europe for commercial video applications. The connector has 21 pins that handle all the input and output connections with one cable.

Scenery:
Generally refers to the furnishings and contents of a film set.

Scratch:
A mark on either a film base or emulsion that can be for either one frame or many.

Screening:
The showing of a completed film or video-generated program material.

SECAM:
An abbreviation of the words "Sequential Couleur avec Memoire," a 625-scan-line system that is used mainly in France.

Secondary Color Correction:
The process of adjusting the parameter of an individual color within the image without affecting the overall color balance of the scene. The basic form of secondary color correction allows the operator to adjust the hue, saturation, and luminance values for red, green, blue, yellow, cyan, or magenta.

Segmented Format:
A format that splits the image into segments, with each on a separate track.

Selective Color Correction:
Masking off a portion of the picture in order to carry out a color correction only in that area. This type of isolation can also be referred to as an area isolation.

Sensitivity:
The degree to which film material reacts to changes.

Sensitometry:
The method of measuring a film emulsion's sensitivity.

Sepia:
A brownish tint seen in many older movies and photographs. It is a wash that is applied over black-and-white material.

Serial Port:
The connection port on a computer that allows communication to external devices using either the RS-232 or RS-422 protocol.

Shuttle:
The method of quickly moving the videotape through the transport.

Side Lighting:
The lighting method that hits the subject from one angle, resulting in shadows that can create a very dramatic effect.

Signal-to-Noise Ratio:
Also known as SNR, this is the level of noise measured against the image information.

Single Perf:
This term refers to film stock that only contains one edge of perforations.

Slate:
A board that is shown at the start of a take and identifies the individual takes. The term also refers to an identification that is recorded after color bars on a videotape, showing the title of the program, date of production, and so on.

SMPTE:
Abbreviation of the Society of Motion Picture and Television Engineers, which is an organization based in the United States.

SMPTE Color Bars:
A color bar test signal introduced by the Society of Motion Picture and Television Engineers. It contains a pluge test signal in the lower part of the test signal.

SMPTE Time Code:
A counting system introduced by SMPTE, the Society of Motion Picture and Television Engineers. It displays a count in hours, minutes, seconds, and frames.

Soft:
The term for an image that may be slightly out of focus.

Specifications:
A listing of requirements intended to ensure that the product meets with the client's required format. Also referred to as specs.

Splicer:
A mechanical device that is used for joining two pieces of film, using either a special glue or splicing tape.

Squeezed:
Refers to an image that has been specially photographed so that it compresses the content of a wide-screen image into a small area, by means of an anamorphic lens. It is later unsqueezed by a special lens when displayed, resulting in a wide-screen image.

Standards Conversion:
A method of changing a program out of one standard into another.

Standby Time:
Waiting time during a session, when work can not proceed. Since this is not considered downtime, it is generally considered billable.

Still Mode:
A mode on a VTR or videotape recorder that shows a still frame when the transport is not in motion.

Still Store:
A unit that records reference frames for later use during a session, to ensure color continuity throughout a session.

Stock:
Generally refers to unused videotape.

Storyboard:
A drawing representing the scenes within a production, intended to give those involved a graphic impression of the project.

Striped Stock:
Videotape material that has black and time code recorded on it, in preparation for insert editing. The term also refers to film stock that has a magnetic sound track placed upon it.

S-VHS:
A format that utilizes a special high-energy videotape to provide resolution superior to that of VHS by receiving the input video in Y/C mode, which separates the chrominance from the luminance information.

S-Video:
A signal connection that has separated chrominance and luminance information. This is also referred to as Y/C, and is used in the S-VHS format.

Sync:
This term refers to the video signal's timing properties, which are vital for showing a video signal.

Tail:
Generally pertains to the end of a reel of film. The beginning of the reel is known as the head.

Tape-to-Tape Color Correction:
The color correction of material that is on videotape and that is recorded onto another videotape, following correction.

TBC:
An abbreviation for "time base corrector."

Technicolor®:
The trademarked motion picture color process.

Telecine:
A device that transfers film material into the video domain.

Telecine Transport Controller:
An external device that controls all the functions of the telecine machine.

Test Pattern:
A test chart that is used as a form of reference during the production and post-production processes.

Theatrical Release Print:
A positive motion picture film print that is made to be shown in a movie theater.

Three-Color Meter:
A device that measures color temperature.

TIFF:
The abbreviation for Tagged Image File Format, a graphics file format initially introduced by Microsoft and Aldus.

Time Code:
A numbering system that is used in video to references each image to a particular number.

Time-Code Character Generator:
A device that visibly generates time-code numbers, generally for the purpose of creating a window dub recording.

Time-Code Window:
A visible window showing time-code information that, when recorded, is referred to as a window dub.

Time Offset:
This term refers to an offset between two different time-code channels.

Timer:
This term refers to either (1) a person in a film laboratory who balances the color of a film negative for positive print reproduction, or (2) a device for measuring the running time of program material.

Tint:
This term refers to either the image's hue or to a wash applied to a particular image, such as a sepia tone wash.

Transparent:
Generally refers to a signal that passes through a device and remains completely unchanged.

Trim:
The method of making a small readjustment to an adjustment already created.

T-Stop:
A number etched on a lens that indicates the opening of the lens's iris.

Uncompressed Video:
Video information that contains absolutely no compression at any stage of the post-production chain. It conforms to the ITU-R 601 standard.

Underexposed:
Pertains to a film exposure that has not received an adequate amount of light.

Underscan:
Term for a video display that shows the entire image. Consumer television monitors display an overscan image that cuts off a small portion around the outer edge of the image.

Unsegmented Format:
A recording format that contains either a complete frame or field on one track.

USB:
Abbreviation of universal serial bus, a method for connecting multiple devices to a computer. These types of ports run at a much faster rate than a normal serial connection.

User Bits:
An allotted area within the time-code information that can contain additional data, such as film keycode, the date, and so on.

VCR:
Short for video cassette recorder.

Vector:
A particular color that is being controlled by a secondary color corrector.

Vectorscope:
The vectorscope displays the hue and saturation properties of the color signal. The display is similar to that of the waveform monitor, with the exception that the measurements are viewed on a circular scale. The outer portion of the circle shows a series of small blocks that indicate the position of red, green, blue, yellow, cyan, and magenta.

Vertical Interval Time Code:
Time code that appears as small white dots in the vertical blanking area of the image. This particular form of time code has the distinct advantage that it will be displayed correctly even if the VTR is not moving. Also known as VITC.

VGA:
Video Graphics Array, an analog graphics display format that can reproduce either 16 or 256 colors.

VHS:
A consumer format that records an analog composite signal along with two channels of audio. The format records using a half-inch-wide videotape enclosed within a specially designed cassette. There are three speeds available: SP, or standard play, which records at 1.94 inches per second, and yields the best results; LP, or long play, which records at 0.66 inches per second; and finally SLP, or super long play, which records at 0.44 inches per second.

VHS-C:
A small version of the VHS format, also referred to as VHS-Compact. It was used basically for camcorders; however, it can be played in conventional machines with the use of a special adapter.

VHS Hi-Fi:
This term refers to an improvement of the audio capabilities within the VHS format.

Video:
This term refers to a signal containing video information.

Videocassette Recorder:
Also known as a VCR, this is a videotape recorder that records using a videotape enclosed in a cassette.

Video Character Generator:
An electronic titling device that is used for creating subtitles, titles, or even identification slates.

Video Color Analyzer:
A device used by a film laboratory color timer for scene-by-scene balancing of the color and density of a motion picture film negative.

Video Frame Delay:
A delay of the video information that is usually caused by the processing within a particular device.

Video Monitor:
This video display device is similar to a consumer television; however, it is usually without a tuner and accommodates only component or composite video signals.

Videotape Master:
A completed videotape that is ready for distribution.

Vignette:
This term generally refers to a soft circular mask around the image. It can be black, or it may contain some color.

VITC:
Short for "vertical interval time code."

VTR:
Videotape recorder.

Waveform Monitor:
An offshoot of the oscilloscope, a device used in laboratories for electronic equipment testing. The waveform monitor exhibits a two-dimensional display of the video signal, using a scale broken down into units.

White Balance:
An adjustment within a video camera that balances the light level. It is achieved by pointing the camera at a chart or card under the lighting to be used for the subject.

Wide-screen:
The wide-screen format is a method for displaying images in a film theater. It can be anything from a 1:85.1 aspect ratio film to a CinemaScope anamorphic film. Wide-screen films fit nicely into the screen format for HDTV, due to the 16 × 9 characteristics of high-definition televisions.

Window Dub:
A special videotape dub that shows the time code visibly within the picture. These are usually made for offline editing purposes.

Wireframe:
An outline display of a particular image.

Work Print:
Typically a one-light motion picture film print that is made for the use of editing. The work print is usually spliced with tape splices, and is generally quite worn due to periodic reediting.

WYSIWYG:
Abbreviation of the words "What You See Is What You Get." It essentially means that what you are observing on the monitor will be the result when it is exported.

X-axis:
The horizontal positioning control.

XGA:
Extended Graphics Array.

Y-axis:
The vertical positioning control.

YCM:
Refers to the three colors yellow, cyan, and magenta.

YCrCb:
A color space, also known as YUV. Y is the luminance information, and Cr and Cb represent the chroma information.

Y, R-Y, B-Y:
The component system used in NTSC. In Y, R-Y, B-Y, the "R" stands for red and the "B" for blue. The "Y" pertains to the luminance information, and the green information is generated from the other components. The use of this method for maintaining isolated signals yields extremely pure colors and excellent sharpness.

Z-axis:
This term generally refers to a Zoom capability.

Index

NON-ALPHABETIC CHARACTERS

A

F

G

H

W

X

Y

Notes and Instructions

This CD-ROM has been configured for compatibility on most Mac and PC Windows operating systems. On a PC, the CD should automatically launch the opening screen in your web browser after it is inserted. If it does not, or if you have closed the browser screen and wish to re-open the CD's start screen, open the window or directory for your CD drive, and double-click on either the STARTUP.HTM or INDEX.HTM file. On a Mac, the CD will not launch automatically. Open the file directory for the CD, and double-click on STARTUP.HTM or INDEX.HTM. If you are prompted to select a program to open the file, select your web browser (Internet Explorer or Netscape).

Please note that in some browsers, the QuickTime™ movie examples will start running in a smaller window. You can adjust the window size for clearer viewing by clicking-and-dragging the window frame sides, and then re-starting the movie.

We would like to acknowledge the depiction of software produced by the following:

Digital Film Tools, LLC © 2002 (Composite Suite Color Correct 2.0®; 55 mm®)

In-Sync Corp. (Speed Razor®; ViXen®)

Adobe (Premiere®; After Effects®)

QuickTime™

The CD-ROM also contains links to websites where trial versions of some of the applications listed above can be downloaded.

Limited Warranty and Disclaimer of Liability

Focal Press and anyone else who has been involved in the creation or production of the accompanying code ("the product") cannot and do not warrant the performance or results that may be obtained by using the product. The product is sold "as is" without warranty of any kind (except as hereafter described), either expressed or implied, including, but not limited to, any warranty of performance or any implied warranty of merchantability or fitness for any particular purpose. Focal Press warrants only that the magnetic CD-ROM(s) on which the code is recorded is free from defects in material and faulty workmanship under the normal use and service for a period of ninety (90) days from the date the product is delivered. The purchaser's sole and exclusive remedy in the event of a defect is expressly limited to either replacement of the CD-ROM(s) or refund of the purchase price, at Focal Press's sole discretion.

In no event, whether as a result of breach of contract, warranty, or tort (including negligence), will Focal Press or anyone who has been involved in the creation or production of the product be liable to purchaser for any damages, including any lost profits, lost savings, or other incidental or consequential damages arising out of the use or inability to use the product or any modifications thereof, or due to the contents of the code, even if focal press has been advised of the possibility of such damages, or for any claim by any other party.

Any request for replacement of a defective CD-ROM must be postage prepaid and must be accompanied by the original defective CD-ROM, your mailing address and telephone number, and proof of date of purchase and purchase price. Send such requests, stating the nature of the problem, to Elsevier Science Customer Service, 6277 Sea Harbor Drive, Orlando, FL 32887, 1-800-321-5068. Focal Press shall have no obligation to refund the purchase price or to replace a CD-ROM based on claims of defects in the nature or operation of the product.

Some states do not allow limitation on how long an implied warranty lasts, nor exclusions or limitations of incidental or consequential damage, so the above limitations and exclusions may not apply to you. This warranty gives you specific legal rights, and you may also have other rights that vary from jurisdiction to jurisdiction.

The re-export of United States origin software is subject to the United States laws under the export administration act of 1969 as amended. Any further sale of the product shall be in compliance with the United States Department of Commerce administration regulations. Compliance with such regulations is your responsibility and not the responsibility of Focal Press.